Math
ADVANTAGE

Take Another Look

RETEACHING WORKBOOK

Harcourt Brace & Company

Orlando • Atlanta • Austin • Boston • San Francisco • Chicago • Dallas • New York • Toronto • London

http://www.hbschool.com

Printed in the United States of America

ISBN 0-15-311064-3

7 8 9 10 11 12 073 06 05 04 03 02 01

CONTENTS

Sums to 10

To find the sum of two numbers you must add.

$$4 + 3 = 7$$
addend   **addend**

Put a counter on each bean to show the addends.
Add. Write the sum.

1.

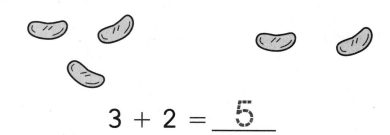

$$3 + 2 = \underline{5}$$

2.

$$7 + 2 = \underline{}$$

3.

$$5 + 3 = \underline{}$$

4.

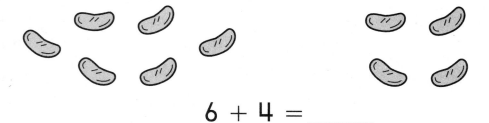

$$6 + 4 = \underline{}$$

Name _____

Order Property

The order of the addends does not change the sum.

$$2 + 4 = \underline{6}$$ $$4 + 2 = \underline{6}$$

Use counters. Write the sums.

1.

$$4 + 5 = \underline{9}$$

$$5 + 4 = \underline{9}$$

2.

$$2 + 7 = \underline{}$$

$$7 + 2 = \underline{}$$

3.

$$3 + 2 = \underline{}$$

$$2 + 3 = \underline{}$$

4.

$$6 + 1 = \underline{}$$

$$1 + 6 = \underline{}$$

5.

$$1 + 9 = \underline{}$$

$$9 + 1 = \underline{}$$

6.

$$1 + 7 = \underline{}$$

$$7 + 1 = \underline{}$$

7.

$$3 + 4 = \underline{}$$

$$4 + 3 = \underline{}$$

8.

$$2 + 6 = \underline{}$$

$$6 + 2 = \underline{}$$

9.

$$8 + 2 = \underline{}$$

$$2 + 8 = \underline{}$$

Name _____

Zero Property

Write the sum.

1.

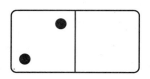

$2 + 0 = \underline{2}$

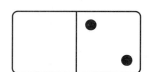

$0 + 2 = \underline{2}$

Any number plus zero equals the same number.

2.

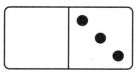

$0 + 3 = \underline{3}$

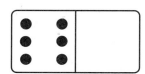

$6 + 0 = \underline{}$

3.

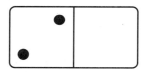

$2 + 0 = \underline{}$

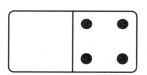

$0 + 4 = \underline{}$

4.

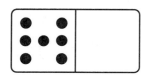

$7 + 0 = \underline{}$

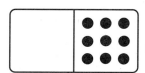

$0 + 9 = \underline{}$

5.

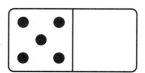

$5 + 0 = \underline{}$

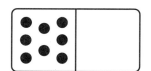

$8 + 0 = \underline{}$

Counting On

Circle the greater number. Count on to add.

1. (4)
      + 2
      ———
        6

     5 , 6

2. 3
      + 5   ——, ——, ——

3. 5
      + 1   ——

4. 1
      + 9   ——

5. 7
      + 2   ——, ——

6. 2
      + 3   ——, ——

7. 3
      + 6   ——, ——, ——

8. 6
      + 1   ——

9. 6
      + 2   ——, ——

Addition Practice

Use ◯ .
Write the sum.

$3 + 2 = \underline{5}$

1. $7 + 1 = \underline{8}$

2. $6 + 3 = \underline{}$

3. $4 + 2 = \underline{}$

4. $8 + 1 = \underline{}$

5. $2 + 0 = \underline{}$

6. $0 + 5 = \underline{}$

7. $9 + 1 = \underline{}$

8. $1 + 4 = \underline{}$

9. $2 + 7 = \underline{}$

10. $3 + 5 = \underline{}$

Differences Through 10

Use connecting cubes. Show the number.
Then take some away. Write how many are left.

$$8 - 2 = \underline{6}$$

	Show	Take Away	How many are left?
1.	8	2	6
2.	7	1	
3.	10	3	
4.	6	4	
5.	4	2	
6.	5	1	
7.	6	3	
8.	10	4	
9.	7	2	
10.	8	6	

Subtracting All or Zero

Use  and a workmat.
Write how many are left.

$6 - 0 = \underline{6}$ $6 - 6 = \underline{0}$

---

1. $5 - 0 = \underline{\hspace{1cm}}$

2. $1 - 1 = \underline{\hspace{1cm}}$

3. $8 - 0 = \underline{\hspace{1cm}}$

4. $9 - 9 = \underline{\hspace{1cm}}$

5. $2 - 0 = \underline{\hspace{1cm}}$

6. $4 - 0 = \underline{\hspace{1cm}}$

7. $3 - 3 = \underline{\hspace{1cm}}$

8. $7 - 0 = \underline{\hspace{1cm}}$

9. $2 - 2 = \underline{\hspace{1cm}}$

10. $5 - 5 = \underline{\hspace{1cm}}$

Using Subtraction to Compare

Use connecting cubes. Compare.
Then subtract.

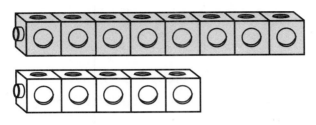

$8 - 5 = \underline{3}$ more

1.

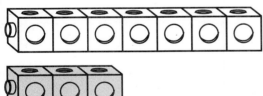

$6 - 4 = \underline{2}$ more

2.

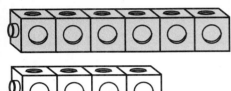

$2 - 1 = \underline{}$ more

3.

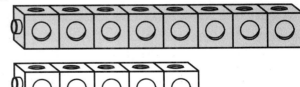

$7 - 3 = \underline{}$ more

4.

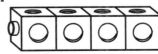

$8 - 5 = \underline{}$ more

5.

$6 - 1 = \underline{}$ more

6.

$4 - 2 = \underline{}$ more

Counting Back

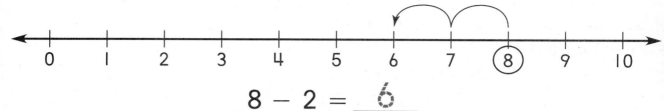

$$8 - 2 = \underline{6}$$

Circle the greater number on the number line.
Draw arrows to count back. Write the difference.

1.

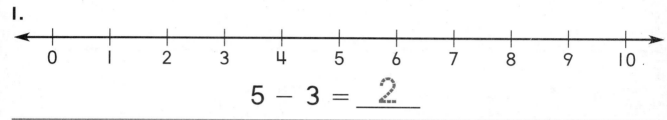

$$5 - 3 = \underline{2}$$

2.

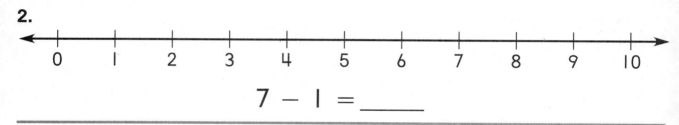

$$7 - 1 = \underline{}$$

3.

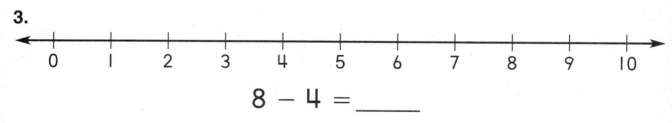

$$8 - 4 = \underline{}$$

4.

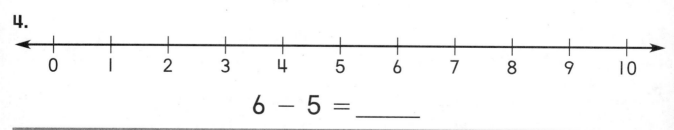

$$6 - 5 = \underline{}$$

5.

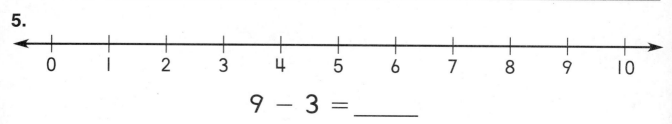

$$9 - 3 = \underline{}$$

Problem Solving • Write a Number Sentence

Draw a line under what you want to know.
Circle the facts. Write a number sentence.
Then write the answer.

1. Sarah walked (2 miles) to school.
   Later she walked home (2 miles.)
   How many miles did she walk?

   $\underline{2}$ $\oplus$ $\underline{2}$ = $\underline{4}$ $\underline{4}$ miles

2. Ben has 5 books about dogs.
   He has 4 books about birds.
   How many books about dogs and birds does he have in all?

   _____ ◯ _____ = _____   _____ books

3. Mary grew 10 flowers in her garden.
   She gave 6 flowers to her grandmother.
   How many flowers does she have left?

   _____ ◯ _____ = _____   _____ flowers

4. Sam had 7 marbles.
   He gave 3 marbles to his sister.
   How many marbles does he have left?

   _____ ◯ _____ = _____   _____ marbles

Doubles

Draw more fruit to show doubles.
Write the addition sentence.

1.

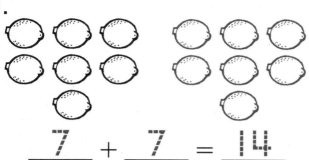

___7___ + ___7___ = ___14___

2.

_____ + _____ = _____

3.

_____ + _____ = _____

4.

_____ + _____ = _____

5.

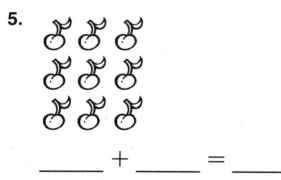

_____ + _____ = _____

6.

_____ + _____ = _____

7.

_____ + _____ = _____

8.

_____ + _____ = _____

More Doubles

○○○ ○○○	○○○ ○○○○	○○○ ○○
3 + 3 = 6	3 + 4 = 7	3 + 2 = 5
doubles	doubles + 1	doubles − 1

Use counters. Write the sum.

	doubles	doubles + 1	doubles − 1
1.	6 + 6 = _12_	6 + 7 = _13_	6 + 5 = _11_
2.	5 + 5 = ____	5 + 6 = ____	5 + 4 = ____
3.	8 + 8 = ____	8 + 9 = ____	8 + 7 = ____
4.	4 + 4 = ____	4 + 5 = ____	4 + 3 = ____
5.	7 + 7 = ____	7 + 8 = ____	7 + 6 = ____
6.	3 + 3 = ____	3 + 4 = ____	3 + 2 = ____
7.	2 + 2 = ____	2 + 3 = ____	2 + 1 = ____

Adding on a Ten-Frame

Use a ten-frame and counters to add 9 + 2.

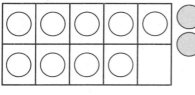

9 + 2 = _____

Move 1 counter to make 10. Add 10 and 1 to make 11.

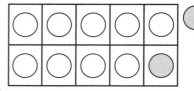

10 + 1 = 11

1.

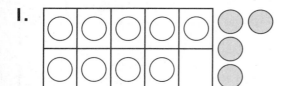

9 + 4 = _____

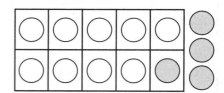

10 + 3 = _13_

2.

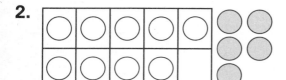

9 + 5 = _____

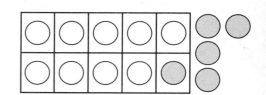

10 + 4 = _____

3.

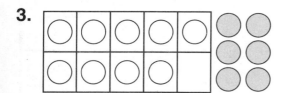

9 + 6 = _____

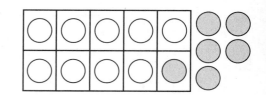

10 + 5 = _____

4.

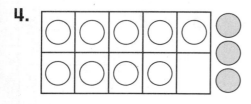

9 + 3 = _____

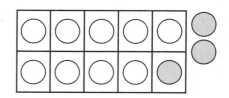

10 + 2 = _____

Make a Ten

Use a ten-frame and counters.

8
+ 6

Move 2 counters to make 10. Add 10 and 4 to make 14.

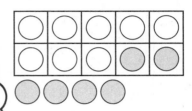

10
+ 4
14

1.

7
+ 4

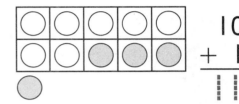

10
+ 1
11

2.

9
+ 6

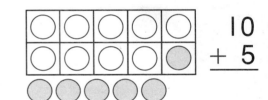

10
+ 5

3.

8
+ 5

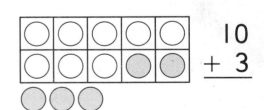

10
+ 3

4.

7
+ 5

10
+ 2

Adding Three Addends

Remember: Make a 10 and use counting on, or use doubles.

(4)
5
+ (4)   >   8
        + 5
        ——
        13

(6)
3
+ (6)   >   12
        + 3
        ——
        15

(9)
8
+ (1)   >   10
        + 8
        ——
        18

1.

(7)
(3)
+ 2
——
12

5
5
+ 4
——

6
4
+ 5
——

4
4
+ 1
——

8
6
+ 2
——

2.

9
1
+ 1
——

6
1
+ 6
——

4
3
+ 3
——

8
2
+ 3
——

9
1
+ 8
——

3.

7
7
+ 2
——

3
7
+ 7
——

6
2
+ 4
——

7
9
+ 1
——

2
5
+ 5
——

Relating Addition and Subtraction

4	+	3	=	7
addend		addend		sum

7	−	3	=	4
sum		addend		difference

Draw more apples to find the sum.
Then cross out apples to find the difference.

I.

9 + 3 = __12__   12 − 3 = __9__

2.

7 + 6 = _____   13 − 6 = _____

3.

8 + 4 = _____   12 − 4 = _____

4.

5 + 5 = _____   10 − 5 = _____

Subtracting on a Number Line

Use the number line to subtract.

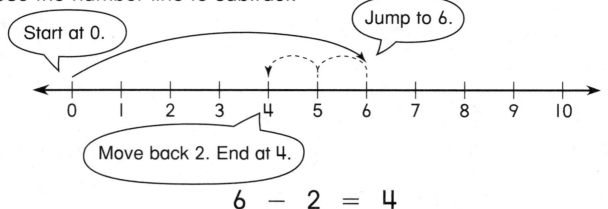

Start at 0.

Jump to 6.

Move back 2. End at 4.

$$6 - 2 = 4$$

1.

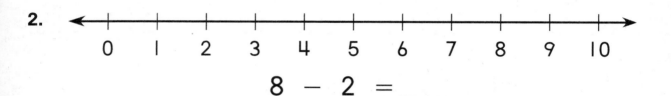

$$10 - 3 = \underline{7}$$

2.

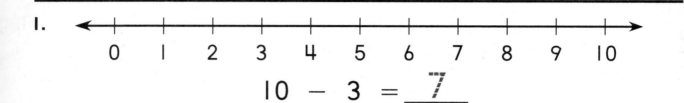

$$8 - 2 = \underline{}$$

3.

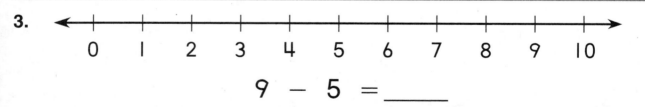

$$9 - 5 = \underline{}$$

4.

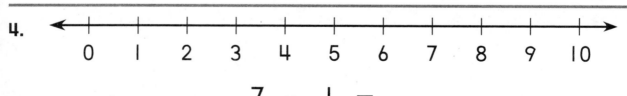

$$7 - 1 = \underline{}$$

5.

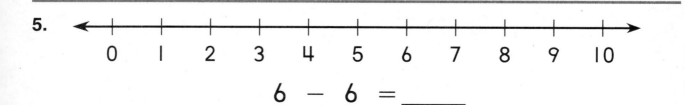

$$6 - 6 = \underline{}$$

R20   **TAKE ANOTHER LOOK**

Name _____

Fact Families

A fact family uses the same numbers in addition and subtraction facts.

$4 + 5 = 9$ $9 - 5 = 4$

$5 + 4 = 9$ $9 - 4 = 5$

2 addition facts   2 subtraction facts

| 4 | 5 | 9 |

numbers in the fact family

Use cubes to show the fact family. Write the sums
and the differences. Write the numbers in the fact family.

1.

$8 + 4 = \underline{12}$ $12 - 4 = \underline{8}$

$4 + 8 = \underline{12}$ $12 - 8 = \underline{4}$

| 8 | 4 | 12 |

2.

$7 + 2 = \underline{\hphantom{00}}$ $9 - 2 = \underline{\hphantom{00}}$

$2 + 7 = \underline{\hphantom{00}}$ $9 - 7 = \underline{\hphantom{00}}$

☐ ☐ ☐

3.

$8 + 6 = \underline{\hphantom{00}}$ $14 - 6 = \underline{\hphantom{00}}$

$6 + 8 = \underline{\hphantom{00}}$ $14 - 8 = \underline{\hphantom{00}}$

☐ ☐ ☐

4.

$8 + 7 = \underline{\hphantom{00}}$ $15 - 7 = \underline{\hphantom{00}}$

$7 + 8 = \underline{\hphantom{00}}$ $15 - 8 = \underline{\hphantom{00}}$

☐ ☐ ☐

Name _____

Missing Addends

$$9 + ? = 16$$

You have 9.

You need 7 more to make 16.

$$9 + 7 = 16$$

Use  to find the missing addend.
Draw them. Complete the number sentence.

1.

$$7 + \underline{} = 15$$

2.

$$9 + \underline{} = 12$$

3.

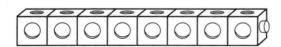

$$8 + \underline{} = 14$$

4.

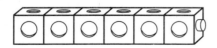

$$6 + \underline{} = 11$$

Name _____

Problem Solving • Choose the Operation

The Red team has 6 baseballs.
The Blue team has 7. How many
baseballs do they have in all?

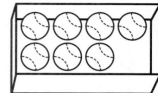

Red Team   **Blue Team**

Add to
join groups.
Use **+**.

Subtract to
compare or take
away. Use **−**.

$$6 \; + \; 7 \; = \; 13$$

Solve. Complete the number sentence.

1. Rob has 8 stickers. Mary has
   6 stickers. How many stickers
   do they have in all?

   $8 \;⊕\; 6 = \underline{14}$ $\underline{14}$ stickers

2. There were 16 balloons
   at the party. 7 popped.
   How many balloons are left?

   $16 \;\bigcirc\; 7 = \underline{\hspace{1cm}}$ $\underline{\hspace{1cm}}$ balloons

3. There are 6 brown puppies. There
   are 4 white puppies. How many
   puppies are there in all?

   $6 \;\bigcirc\; 4 = \underline{\hspace{1cm}}$ $\underline{\hspace{1cm}}$ puppies

4. I counted 7 baseballs and 9 footballs.
   How many more footballs are there
   than baseballs?

   $9 \;\bigcirc\; 7 = \underline{\hspace{1cm}}$ $\underline{\hspace{1cm}}$ more
   footballs

TAKE ANOTHER LOOK   R23

Name _____

Grouping Tens

Use cubes. Make each group of ten.
Count by tens.
Then write how many ones.

ten ones = 1 ten

1.

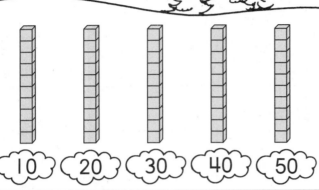

5 tens = __50__ ones

2.

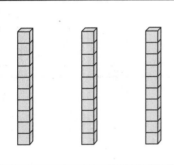

3 tens = _____ ones

3.

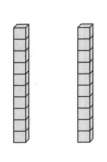

2 tens = _____ ones

4.

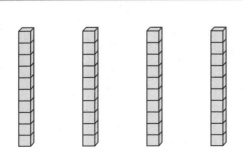

4 tens = _____ ones

Tens and Ones to 50

Show.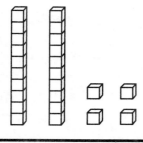

There are 2 tens
and 4 ones.
There are 24 in all.

Write ___24___ .

Use tens and ones blocks to show the model.
Then write the number.

1.

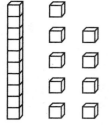

1 ten   9 ones = ___19___

2.

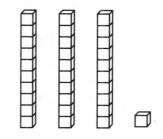

3 tens   2 ones = _____

3.

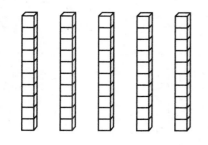

5 tens   0 ones _____

4.

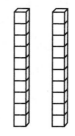

2 tens   0 ones = _____

5.

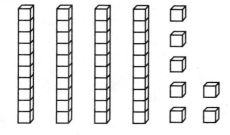

4 tens   7 ones = _____

6.

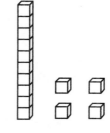

1 ten   4 ones = _____

Tens and Ones to 100

Remember to count your base-ten blocks carefully to tell how many.

There are **3** tens and **4** ones. There are 34 in all.

There are **4** tens and **3** ones. There are 43 in all.

Write how many tens and ones. Circle the number.
Then use base-ten blocks to check.

1.

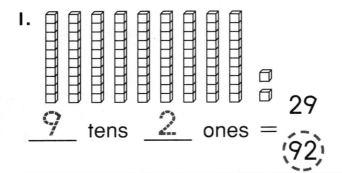

___9___ tens ___2___ ones =

29

(92)

2.

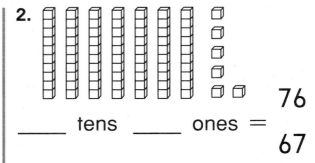

_____ tens _____ ones =

76

67

3.

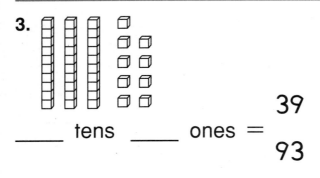

_____ tens _____ ones =

39

93

4.

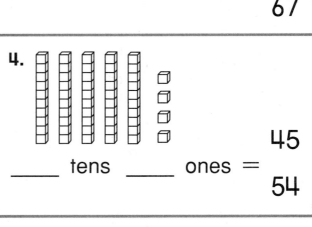

_____ tens _____ ones =

45

54

5.

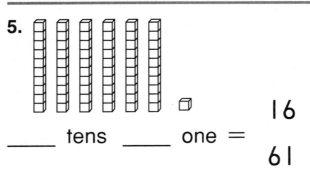

_____ tens _____ one =

16

61

6.

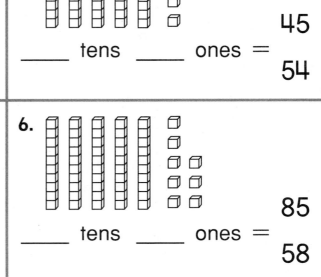

_____ tens _____ ones =

85

58

Use a Model

Use base-ten blocks. Show the tens and ones.
Then write the number.

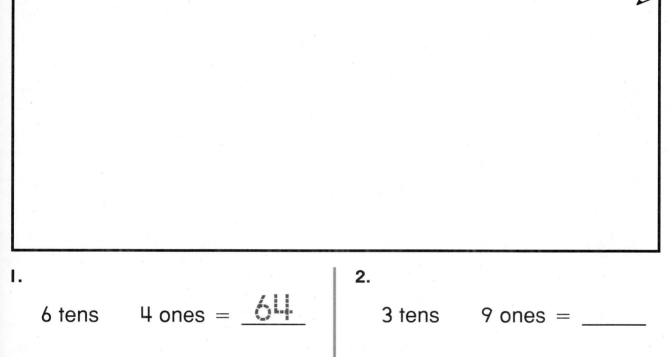

1.

6 tens   4 ones = _64_

2.

3 tens   9 ones = _____

3.

0 tens   9 ones = _____

4.

2 tens   5 ones = _____

5.

3 tens   7 ones = _____

6.

1 ten   8 ones = _____

7.

8 tens   0 ones = _____

8.

4 tens   4 ones = _____

9.

7 tens   6 ones = _____

10.

5 tens   2 ones = _____

Exploring Estimation

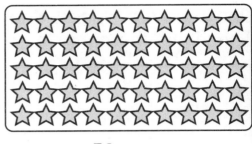

| 10 stars | 25 stars | 50 stars |

Look at each group of stars. Use these groups
to help you choose the better estimate.

1.

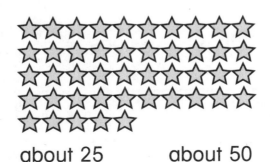

about 10   (about 25)

2.

about 10   about 25

3.

about 10   about 25

4.

about 25   about 50

5.

about 25   about 50

6.

about 10   about 25

Skip-Counting by Fives and Tens

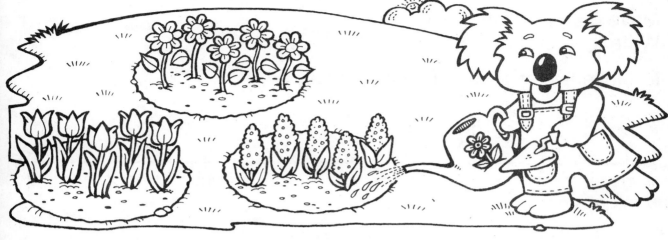

Cut out the cards. Paste them where they
will complete a pattern.

1.

| 5 | 10 | 15 | 20 | |

2.

| 10 | 20 | 30 | | 50 |

3.

 10   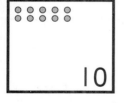 15   [ ]   [ ]   30

| 25 | 20 | 25 | 40 |

Skip-Counting by Twos and Threes

Move a ▮ red ▷ on the number line.
Write the missing numbers.

1. Move two spaces to count by twos.

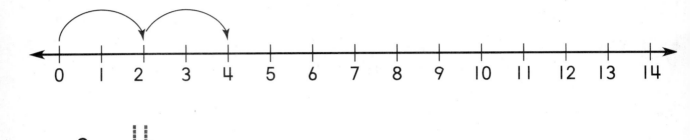

2, _____4_____, _____, _____, _____, _____, _____

2. Move three spaces to count by threes.

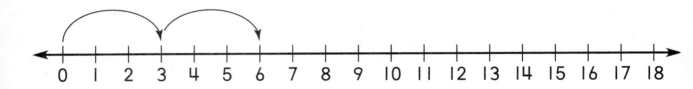

3, _____, _____, _____, _____, _____

3. Move two spaces to count by twos.

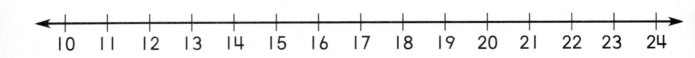

10, _____, _____, _____, _____, _____, _____, _____

Even and Odd Numbers

Circle pairs of dots. Look for dots left over.

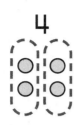

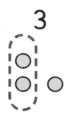

Even numbers have none left over. | **Odd** numbers have 1 left over.

The number 4 is **even.** | The number 3 is **odd.**

Circle pairs of dots. Write **even** or **odd.**

1.

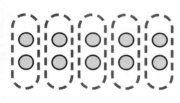

_____**even**_____

2.

_____

3.

_____

4.

_____

5.

_____

6.

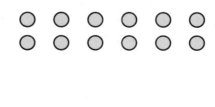

_____

Counting On and Back by Tens

Use tens and ones to show the pattern. Write the number.

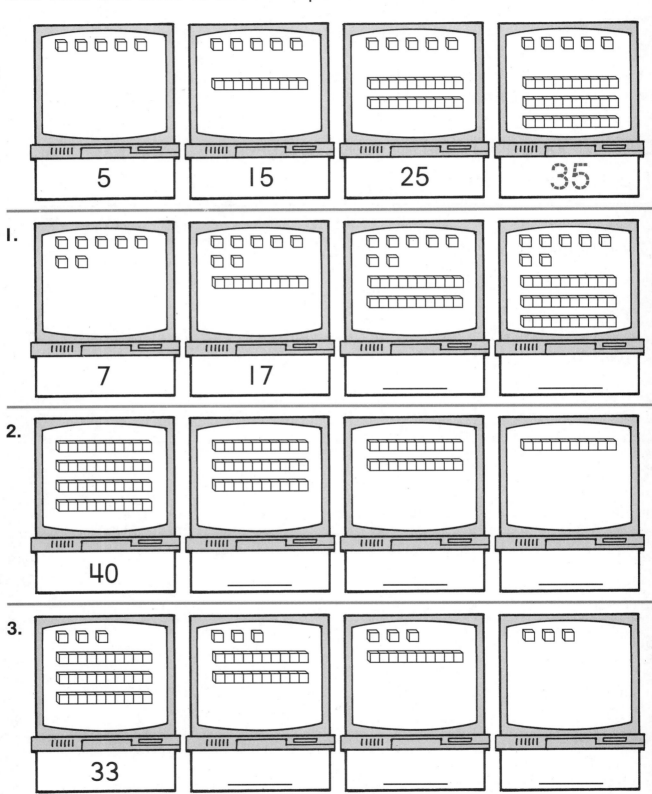

5   **15**   **25**   **35**

1.   **7**   **17**   _____   _____

2.   **40**   _____   _____   _____

3.   **33**   _____   _____   _____

Name _____

Problem Solving • Look for a Pattern

Look at the rule. Draw pictures to complete
the pattern. Write the missing numbers.

1. Count by twos.

2   4   6   8   10   12

2. Count by threes.

3   6   9   ___   ___

3. Count by tens.

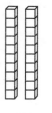 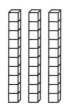 

20   30   40   ___   ___

4. Count by fives.

5   10   15   ___   ___

Comparing Numbers

Use tens and ones to show each number.
Circle the number that is greater.

I. 11	(16)	**2.** 35	32
3. 51	48	**4.** 24	29
5. 39	65	**6.** 19	21

Use tens and ones to show each number.
Circle the number that is less.

7. 10	15	**8.** 27	24
9. 48	47	**10.** 37	41
11. 66	76	**12.** 18	28

Greater Than and Less Than

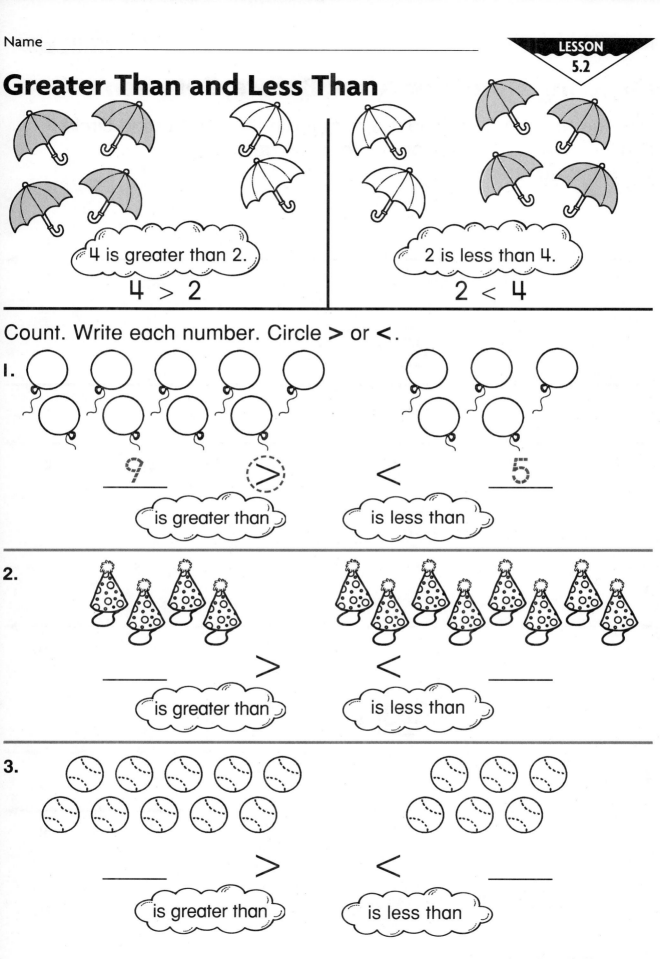

4 is greater than 2.

4 > 2

2 is less than 4.

2 < 4

Count. Write each number. Circle > or <.

1. 9 ___ ⬭> ___ < ___ 5

is greater than   is less than

2. ___ > ___   ___ < ___

is greater than   is less than

3. ___ > ___   ___ < ___

is greater than   is less than

Ordering Numbers: After, Before, Between

Use the number line. Find the number in the square and circle it.
Draw ⌒↘ to show **after** or ↙⌒ to show **before.**
Write the number.

1.

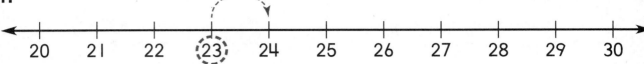

The number after 23 is ___24___ .

2.

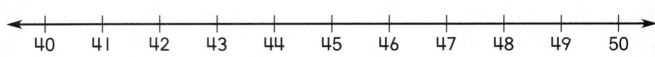

The number after 46 is _____ .

3.

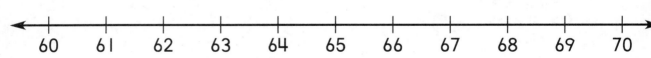

The number before 63 is _____ .

4.

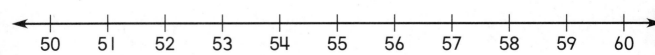

The number before 57 is _____ .

Ordinal Numbers

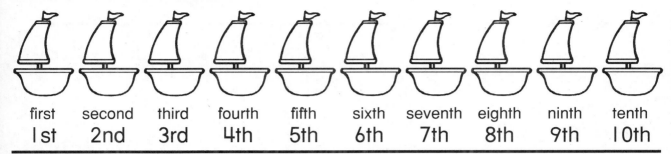

| first | second | third | fourth | fifth | sixth | seventh | eighth | ninth | tenth |
| 1st | 2nd | 3rd | 4th | 5th | 6th | 7th | 8th | 9th | 10th |

Color to show the order.

1. seventh

2. ninth

3. first

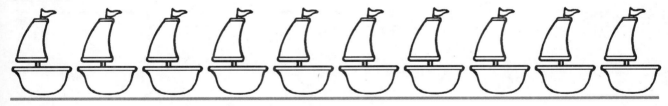

4. fourth

5. sixth

Using a Number Line to Estimate

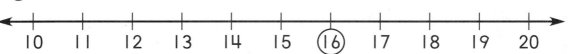

Count. There are 4 spaces from 16 to 20.
There are 6 spaces from 10 to 16.
16 is closer to 20 than 10.

Find the number in the square on the number line.
Circle the answer that tells which ten it is closer to.

1. 18

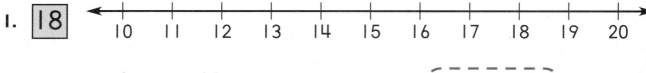

closer to 10   closer to 20

---

2. 23

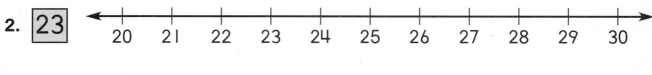

closer to 20   closer to 30

---

3. 39

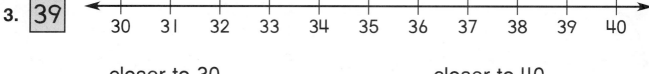

closer to 30   closer to 40

---

4. 44

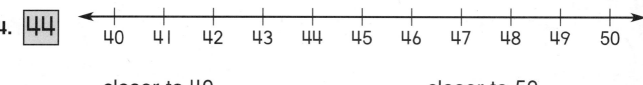

closer to 40   closer to 50

---

5. 27

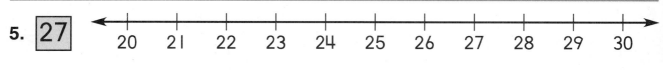

closer to 20   closer to 30

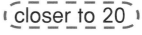

Pennies, Nickels, and Dimes

10¢   5¢   1¢

Begin with the coin of greatest value when you count. 10¢ and 5¢ is 15¢ and 1¢ more is 16¢.

Cover the coins with punch-out coins.
Count on to find the total amount.

1.

10 ¢, 15 ¢, 16 ¢, 17 ¢, 18 ¢ | 18 |¢

2.

_____ ¢, _____ ¢, _____ ¢, _____ ¢, _____ ¢ | |¢

3.

_____ ¢, _____ ¢, _____ ¢, _____ ¢, _____ ¢ | |¢

4.

_____ ¢, _____ ¢, _____ ¢, _____ ¢ | |¢

Name _____

Nickels, Dimes, and Quarters

25¢   **10¢**   **5¢**

Begin with
the coin of greatest
value when you count.
25¢ and 10¢ is 35¢
and 5¢ more
is 40¢.

Cover the coins with punch-out coins.
Count on to find the total amount.

1.

__25__ ¢, __35__ ¢, __40__ ¢, __41__ ¢   **41** ¢

2.

_____ ¢, _____ ¢, _____ ¢, _____ ¢   [ ] ¢

3.

_____ ¢, _____ ¢, _____ ¢, _____ ¢   [ ] ¢

4.

_____ ¢, _____ ¢, _____ ¢, _____ ¢   [ ] ¢

5.

_____ ¢, _____ ¢, _____ ¢, _____ ¢   [ ] ¢

Counting Collections

Cover the coins with punch-out coins.
Then put the punch-out coins in order
from greatest to least. Count the money.
Write the total amount.

1.

$$25¢ \quad 10¢ \quad 5¢$$

$$1¢ \quad 1¢ \quad \underline{42} \; ¢$$

2.

_____ ¢

3.

_____ ¢

4.

_____ ¢

Counting Half-Dollars

There are many ways to make 25¢.
Here are three ways.

There are even more ways to make 50¢.
Count the money. Write the total.
Circle the coins if they make 50¢.

1.

_35_ ¢

2.

_50_ ¢

3.

_____ ¢

4.

_____ ¢

5.

_____ ¢

6.

_____ ¢

Problem Solving • Act It Out

Circle the group of coins that show the price.

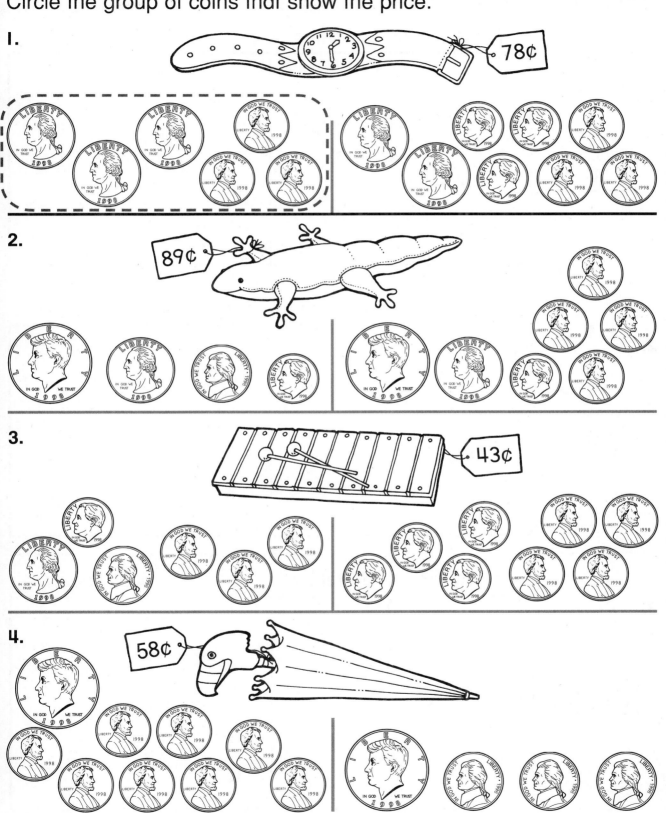

1. 78¢

2. 89¢

3. 43¢

4. 58¢

Combinations of Coins

Use coins. Show the amount in two ways.

	half-dollar and pennies	quarters and pennies
1. 	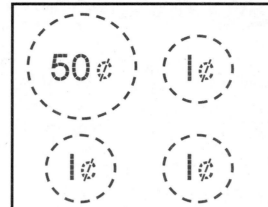	
2. 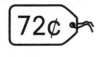	half-dollar, dimes, pennies	quarters, dimes, pennies
3. 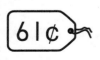	half-dollar, dime, penny	quarters, dime, penny

Equal Amounts Using Fewest Coins

Show the amount with fewer coins.

1.

 75¢

Trade 2 quarters for a half-dollar. Trade 2 dimes and a nickel for a quarter.

2.

 80¢

3.

 45¢

4.

 56¢

Name _____

Comparing Amounts to Prices

Circle the coins you need to buy the toy.

1.
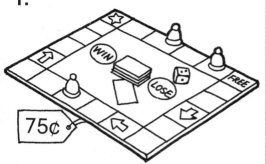 75¢

2.
 69¢

3.
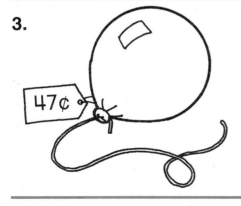 47¢

4.
 55¢

R46   TAKE ANOTHER LOOK

Making Change

You count on to make change. Start with the price
of the item you buy. Count on to the amount you pay.
Find the change for each item.

1. You pay 30¢.

27¢

Count on from
27¢ to 30¢.

<u>28</u> ¢, <u>29</u> ¢, <u>30</u> ¢

Your change is <u>3</u> ¢.

2. You pay 45¢.

Count on from
41¢ to 45¢.

41¢

_____ ¢, _____ ¢, _____ ¢, _____ ¢

Your change is _____ ¢.

3. You pay 50¢.

Count on from
46¢ to 50¢.

46¢

_____ ¢, _____ ¢, _____ ¢, _____ ¢

Your change is _____ ¢.

4. You pay 75¢.

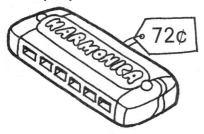

72¢

Count on from
72¢ to 75¢.

_____ ¢, _____ ¢, _____ ¢

Your change is _____ ¢.

Problem Solving • Act It Out

Circle the coins you need to buy the object.
If you do not have enough money,
circle **not enough money**.

1.

not enough money

2.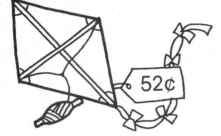

not enough money

3.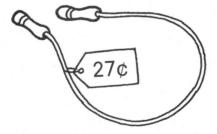

not enough money

4.

not enough money

5.

not enough money

Hour and Half-Hour

These two clocks show 4:00.

These two clocks show 4:30.

Read the time. Choose the matching time from the box
to write the time on the second clock.

11:30	1:00	10:30	7:30	9:00

1.

 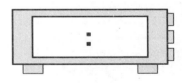

2.

 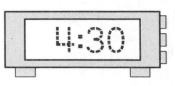

3.

 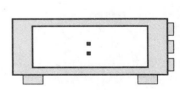

4.

5.

 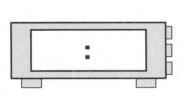

6.

 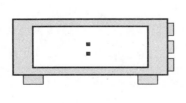

TAKE ANOTHER LOOK R49

Telling Time to 5 Minutes

This clock shows 5 o'clock.

5:00

00   05

This clock shows 5 minutes after 5 o'clock.

5:05

Use a punch-out clock.
Write the time.

1.

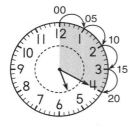

05   10

5:10

2.

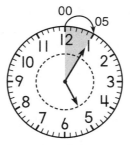

00   05   10   15

5:___

3.

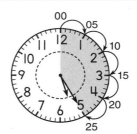

00   05   10   15   20

5:___

4.

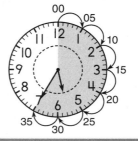

00   05   10   15   20   25

5:___

5.

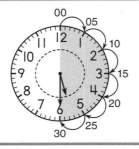

00   05   10   15   20   25   30

5:___

6.

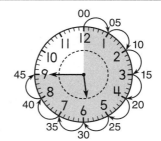

00   05   10   15   20   25   30   35

5:___

7.

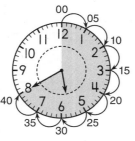

00   05   10   15   20   25   30   35   40

5:___

8.

45   00   05   10   15   20   25   30   35   40

5:___

Name _____

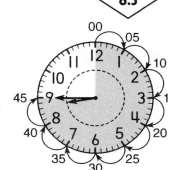

Telling Time to 15 Minutes

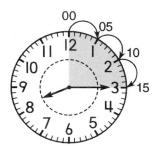

This clock shows the time is 8:15.

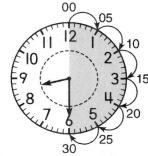

This clock shows the time is 8:30.

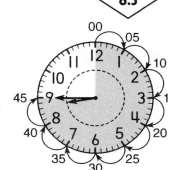

This clock shows the time is 8:45.

Write the time.

I.

5:00

___:___

___:___

2.

___:___

___:___

___:___

3.

___:___

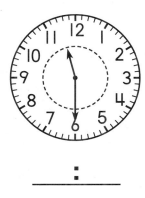

___:___

___:___

Practice Telling Time

Circle the clock that shows the time in the square.

1.

9:15

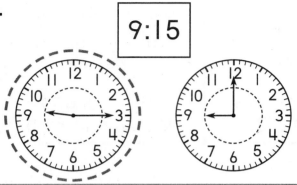

2.

7:10

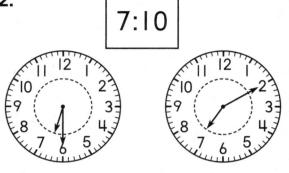

3.

11:40

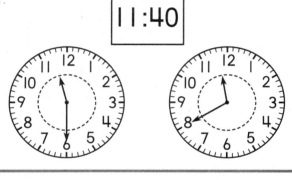

4.

3:00

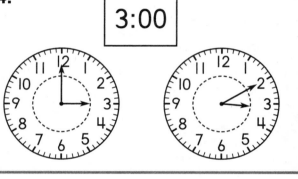

5.

5:30

6.

6:35

7.

12:05

8.

11:40

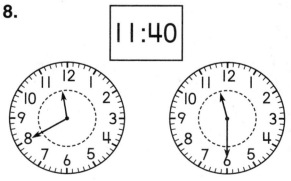

Elapsed Time

Cut out the clocks. Find the clock that shows the new time. Paste it in the box.

Count on 2 hours to find the new time

1.

The party started at ____.

It will last 2 hours. What time will it be over?

2.

Beth started walking at ____.

She walked for 30 minutes. What time did she get home?

Paste here.

3.

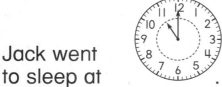

Jack went to sleep at ____.

He slept for 1 hour. What time did he wake up?

Paste here.

4.

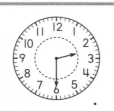

Tasha went swimming at ____.

She swam for 30 minutes. What time did she stop?

Paste here.

5.

Josh went to the library at ____.

He stayed for 2 hours. What time did he come home?

Paste here.

6.

Tyrone went to the park at ____.

He played for 30 minutes. What time did he leave the park?

Paste here.

 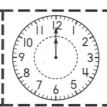 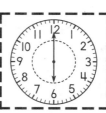  

Name _____

Reading a Calendar

Use the calendar to answer the questions.

July

Sunday	Monday	Tuesday	Wednesday	Thursday	Friday	Saturday
	1	2	3	4 Independence Day	5	6
7	8	9	10	11	12	13
14	15	16	17	18	19	20
21	22	23	24	25	26	27
28	29	30	31			

1. Mark an X on every Friday.

2. Circle the name of the month.

3. Draw a ☆ on Independence Day.

4. Draw a ～ on the second week of the month.

5. Circle the fourth Tuesday of the month.

6. Draw a △ on the last day of the month.

Using a Calendar

January

S	M	T	W	T	F	S
				1	2	3	4
5	6	7	8	9	10	11
12	13	14	15	16	17	18
19	20	21	22	23	24	25
26	27	28	29	30	31	

February

S	M	T	W	T	F	S
						1
2	3	4	5	6	7	8
9	10	11	12	13	14	15
16	17	18	19	20	21	22
23	24	25	26	27	28	

March

S	M	T	W	T	F	S
						1
2	3	4	5	6	7	8
9	10	11	12	13	14	15
16	17	18	19	20	21	22
23/30	24/31	25	26	27	28	29

April

S	M	T	W	T	F	S
		1	2	3	4	5
6	7	8	9	10	11	12
13	14	15	16	17	18	19
20	21	22	23	24	25	26
27	28	29	30			

May

S	M	T	W	T	F	S
				1	2	3
4	5	6	7	8	9	10
11	12	13	14	15	16	17
18	19	20	21	22	23	24
25	26	27	28	29	30	31

June

S	M	T	W	T	F	S
1	2	3	4	5	6	7
8	9	10	11	12	13	14
15	16	17	18	19	20	21
22	23	24	25	26	27	28
29	30					

July

S	M	T	W	T	F	S
		1	2	3	4	5
6	7	8	9	10	11	12
13	14	15	16	17	18	19
20	21	22	23	24	25	26
27	28	29	30	31		

August

S	M	T	W	T	F	S
					1	2
3	4	5	6	7	8	9
10	11	12	13	14	15	16
17	18	19	20	21	22	23
24/31	25	26	27	28	29	30

September

S	M	T	W	T	F	S
	1	2	3	4	5	6
7	8	9	10	11	12	13
14	15	16	17	18	19	20
21	22	23	24	25	26	27
28	29	30				

October

S	M	T	W	T	F	S
			1	2	3	4
5	6	7	8	9	10	11
12	13	14	15	16	17	18
19	20	21	22	23	24	25
26	27	28	29	30	31	

November

S	M	T	W	T	F	S
						1
2	3	4	5	6	7	8
9	10	11	12	13	14	15
16	17	18	19	20	21	22
23/30	24	25	26	27	28	29

December

S	M	T	W	T	F	S
	1	2	3	4	5	6
7	8	9	10	11	12	13
14	15	16	17	18	19	20
21	22	23	24	25	26	27
28	29	30	31			

Circle the answer.

1. The first month of the year. **red**
2. The name of the month that follows April. **blue**
3. The seventh month. **green**
4. The month with the fewest days. **purple**
5. Your birthday month. **orange**
6. The months that have 31 days. **brown**

TAKE ANOTHER LOOK   R55

Early or Late

Use the clocks. Circle early or late.

1. Swimming starts at 2:30.
Is Elena early or late?

(early)

late

---

2. Marcie's party starts at 4:00.
Is Joe early or late?

early

late

---

3. Jake is meeting Mike at 9:15.
Is Jake early or late?

early

late

9:30

---

4. The school play starts at 7:00.
Is Ed early or late?

early

late

6:45

Sequencing Events

Jan has to wash her dog.
Cut and paste the events in order.

1.

2.

3.

4.

Problem Solving • Reading a Schedule

Look at the times. Fill in the schedule.

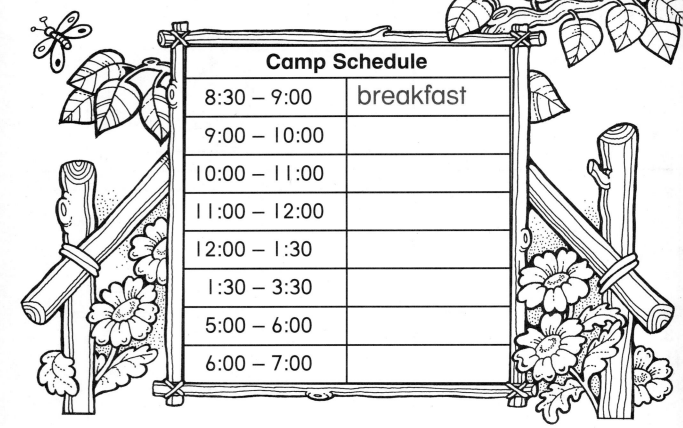

Camp Schedule	
8:30 – 9:00	breakfast
9:00 – 10:00	
10:00 – 11:00	
11:00 – 12:00	
12:00 – 1:30	
1:30 – 3:30	
5:00 – 6:00	
6:00 – 7:00	

1. Breakfast is from 8:30 – 9:00.
2. Lunch is from 11:00 – 12:00.

3. Swimming is from 1:30 – 3:30.
4. Games are from 6:00 – 7:00.

5. Sports are from 9:00 – 10:00.
6. Dinner is from 5:00 – 6:00.

7. Crafts are from 10:00 – 11:00.
8. Rest is from 12:00 – 1:30.

Regrouping Ones as Tens

Count. Can you make a ten? If so, regroup 10 ones as 1 ten.

14 ones

tens	ones
	☐ ☐ ☐ ☐ ☐ ☐ ☐ ☐ ☐ ☐ ☐ ☐

tens	ones
▭▭▭▭▭	☐ ☐ ☐ ☐ ☐ ☐ ☐ ☐ ☐ ☐ ☐ ☐ ☐ ☐

1 ten 4 ones

tens	ones
▭▭▭▭▭▭	☐ ☐ ☐ ☐

Use Workmat 3 and base-ten blocks.
Use ones to show the number.
Regroup if you need to. Write how many tens and ones.

1.
14 ones = __1__ ten __4__ ones

2.
7 ones = ____ tens ____ ones

3.
18 ones = ____ ten ____ ones

4.
13 ones = ____ ten ____ ones

5.
15 ones = ____ ten ____ ones

6.
10 ones = ____ ten ____ ones

7.
12 ones = ____ ten ____ ones

8.
9 ones = ____ tens ____ ones

9.
17 ones = ____ ten ____ ones

10.
11 ones = ____ ten ____ one

Modeling One-Digit and Two-Digit Addition

1.

$12 + 9 =$ _____

Show 12.
Then show 9.

Join the ones.
Can you make a ten?

Yes   No

If so, regroup
10 ones as 1 ten.

Write how many
in all.

21

tens	ones

tens	ones

tens	ones

Use Workmat 3 and base-ten blocks.

Show.	Join the ones. Can you make a ten? If so, regroup 10 ones as 1 ten.		Write how many in all.
2. 11 + 6	Yes	No	17
3. 9 + 13	Yes	No	_____
4. 12 + 4	Yes	No	_____
5. 9 + 17	Yes	No	_____
6. 18 + 5	Yes	No	_____

Modeling Two-Digit Addition

1.

$13 + 29 =$ ___

Show 13.
Then show 29.

Join the ones.
Can you make a ten?

(Yes)   No

If so, regroup
10 ones as 1 ten.

Write how many
in all.

42

tens	ones

tens	ones

tens	ones

Use Workmat 3 and base-ten blocks.

Show.	Can you make a ten? If so, regroup 10 ones as 1 ten.		Write how many in all.
2. 12 + 16	Yes	(No)	28
3. 15 + 19	Yes	No	___
4. 18 + 13	Yes	No	___
5. 11 + 14	Yes	No	___
6. 16 + 17	Yes	No	___
7. 11 + 13	Yes	No	___

Recording Two-Digit Addition

1.

tens	ones
1	4
+ 3	8
5	2

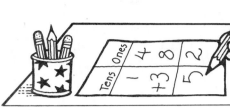

Show 14.
Then show 38.

Join the ones.
Regroup if you need to.

Add the tens.
Write how many.

tens	ones

tens	ones

tens	ones

2.

tens	ones
2	9
+ 1	8

tens	ones

3.

tens	ones
4	8
+ 2	4

tens	ones

4.

tens	ones
3	2
+ 1	6

tens	ones

5.

tens	ones
2	6
+ 3	5

tens	ones

Problem Solving • Make a Model

tens	ones
1	2
+	8
2	0

I. There are 12 girls and
8 boys in the play.
How many children are in the play?

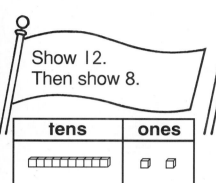

Show 12.
Then show 8.

tens	ones

Join the ones.
Regroup if you need to.

tens	ones

Add the tens.
Write how many.

tens	ones

Use Workmat 3 and base-ten blocks.

2. There are 11 boys and
13 girls playing in the park.
How many children are
playing in the park?

_____ children

tens	ones
1	1
+ 1	3

---

3. There are 15 boys and
16 girls at the picnic.
How many children are
at the picnic?

_____ children

tens	ones
1	5
+ 1	6

---

4. There are 18 girls and
9 boys playing on the
playground. How many
children are playing on
the playground?

_____ children

tens	ones
1	8
+	9

Adding One-Digit and Two-Digit Numbers

Use base-ten blocks and Workmat 3.

Step 1
Add the ones.
$7 + 6 = 13.$

Step 2
Regroup 13 ones to make 1 ten and 3 ones. Write 1 ten to show the new ten.

Step 3
Add the tens. Write how many.

tens	ones

tens	ones
3	7
+	6

tens	ones

tens	ones
3	7
+	6
	3

tens	ones

tens	ones
3	7
+	6
4	3

1.

tens	ones
5	1
+	9

tens	ones

2.

tens	ones
3	6
+	3

tens	ones

3.

tens	ones
6	7
+	7

tens	ones

4.

tens	ones
7	6
+	8

tens	ones

Adding Two-Digit Numbers

Use base-ten blocks and Workmat 3.
Add. Do you need to regroup?
Circle the workmat that shows the correct answer.

1.

tens	ones
3	9
+4	3

(Yes)

No

tens	ones

tens	ones

2.

tens	ones
5	2
+1	9

Yes

No

tens	ones

tens	ones

3.

tens	ones
6	3
+2	6

Yes

No

tens	ones

tens	ones

4.

tens	ones
7	4
+1	7

Yes

No

tens	ones

tens	ones

More About Two-Digit Addition

Use base-ten blocks and Workmat 3.

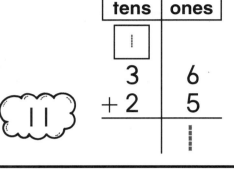

Add the ones. Write how many.

Regroup if you need to. Write the tens and ones.

Add the tens.

tens	ones
□	
3	6
+2	5

| | | |

tens	ones
¦	
3	6
+2	5
	¦

tens	ones
¦	
3	6
+2	5
6	1

1.

tens	ones
¦	
2	6
+2	7
5	3

13

tens	ones
□	
3	9
+1	7

tens	ones
□	
4	5
+1	4

2.

tens	ones
□	
5	3
+1	7

tens	ones
□	
7	2
+1	7

tens	ones
□	
6	3
+2	8

3.

tens	ones
□	
3	5
+	7

tens	ones
□	
7	6
+1	4

tens	ones
□	
1	3
+	6

Addition Practice

Add the
ones.

49
+ 23

9 + 3 = 12

Regroup
if you need to.

49
+ 23
2

Add the
tens.

49
+ 23
72

1.

58
+ 7
65

24
+ 36

33
+ 3

66
+ 15

2.

47
+ 26

35
+ 19

58
+ 13

29
+ 9

3.

14
+ 9

27
+ 5

38
+ 24

41
+ 17

4.

73
+ 18

52
+ 9

43
+ 17

65
+ 25

5.

58
+ 8

25
+ 35

69
+ 17

74
+ 16

Problem Solving • Too Much Information

Draw a line through the sentence that is not needed.
Then solve the problem that uses the correct information.

1. Jeremy ate 13 crackers on
   Tuesday. He ate 17 on Thursday.
   ~~He ate 5 apples that week.~~ How
   many crackers did he eat in all?

   _**30**_ crackers

$$\begin{array}{r} 1 \\ 13 \\ +\ 17 \\ \hline 30 \end{array}$$

2. Sally saw 26 goldfish in the
   pond. Her dad put in 36 more
   goldfish. The pond had 6 ducks
   on it. How many goldfish
   were in Sally's pond?

   _____ goldfish

3. Kate planted 45 pansies.
   She planted 16 petunias.
   She made 10 rows in the garden.
   How many flowers did she plant?

   _____ flowers

4. Mr. Jones's class hunted shells.
   The boys found 46. The girls
   found 25. A man at the beach
   found 62. How many shells did
   Mr. Jones's class find?

   _____ shells

Name _____

Regrouping
Tens as Ones

1. Show 2 tens and 4 ones. Are there enough ones to subtract 4 ones?

Yes   No

tens	ones

If not, regroup 1 ten as 10 ones.

tens	ones

Subtract the ones. Write how many tens and ones are left.

1   6
ten   ones

tens	ones

Use Workmat 3 and base-ten blocks. Subtract.

Show the tens and ones.	Subtract.	Are there enough ones to subtract? If not, regroup 1 ten as 10 ones.	Subtract the ones. Write how many tens and ones are left.
2. 2 tens 8 ones	9 ones	Yes **No**	_1_ ten _9_ ones
3. 3 tens 6 ones	8 ones	Yes No	____ tens ____ ones
4. 2 tens 5 ones	7 ones	Yes No	____ ten ____ ones
5. 4 tens 8 ones	6 ones	Yes No	____ tens ____ ones

Modeling One-Digit and Two-Digit Subtraction

1. $37 - 8 =$ _____

Show 37. Are there enough ones to subtract 8 ones?

Yes   (No)

tens	ones

Do you need to regroup?

(Yes)   No

If so, regroup 1 ten as 10 ones.

tens	ones

Subtract the ones. Write how many are left.

$37 - 8 =$ 29

tens	ones

Use base-ten blocks and Workmat 3. Subtract.

Subtract.	Do you need to regroup?		Write how many are left.
2. $43 - 6$	(Yes)	No	37
3. $25 - 6$	Yes	No	_____
4. $35 - 5$	Yes	No	_____
5. $22 - 7$	Yes	No	_____
6. $37 - 9$	Yes	No	_____
7. $14 - 5$	Yes	No	

Recording Subtraction

tens	ones
4	3
−	6
3	7

Show 43.
Subtract 6.

Regroup if
you need to.

Subtract the ones.
Write how many
tens and ones
are left.

tens	ones

tens	ones

tens	ones

Use Workmat 3 and base-ten blocks.
Write how many tens and ones are left.

1.

tens	ones
3	3
−	6
2	7

2.

tens	ones
2	6
−	8

3.

tens	ones
3	2
−	7

4.

tens	ones
2	1
−	5

5.

tens	ones
2	2
−	4

6.

tens	ones
1	5
−	9

7.

tens	ones
4	6
−	8

8.

tens	ones
3	4
−	7

Recording Two-Digit Subtraction

tens	ones
3	4
− 1	8
1	6

Show 34. Are there enough ones to subtract 8? Regroup 1 ten as 10 ones.

Subtract the ones. Write how many are left.

Subtract the tens. Write how many are left.

tens	ones

tens	ones

tens	ones

1.

tens	ones
4	3
− 1	6
2	7

2.

tens	ones
5	6
− 2	7

3.

tens	ones
2	3
− 1	4

4.

tens	ones
3	5
− 2	9

5.

tens	ones
5	4
− 4	5

6.

tens	ones
3	2
− 1	5

7.

tens	ones
2	5
− 1	8

8.

tens	ones
4	8
− 3	9

Problem Solving • Choose the Operation

How many are there **in all**?	How many are **left**?
$5 + 2 = 7$	$7 - 5 = 2$
There are 7 **in all**.	There are 2 **left**.

Circle **add** or **subtract**.
Write + or −.
Find the sum or difference.

1. Pete has 34 marbles.
He gives 15 marbles to Juan.
How many marbles does Pete have left?

add   (subtract)

tens	ones
3	4
⊝ 1	5
1	9

__19__ marbles

2. Mike counts 14 kites.
Then he counts 18 more.
How many kites did Mike count in all?

add   subtract

tens	ones
1	4
◯ 1	8

_____ kites

3. Erin has 28 cookies.
She shares 19 of them with her friends.
How many cookies are left?

add   subtract

tens	ones
2	8
◯ 1	9

_____ cookies

Subtracting One-Digit from Two-Digit Numbers

Use Workmat 3 and base-ten blocks.
Subtract. Circle **Yes** or **No** if you need to regroup.

Step I
Show 35. Do you need to regroup?

Step 2
Regroup I ten as 10 ones. Now there are 15 ones. Subtract 7 from 15. Write how many ones are left.

Step 3
Write how many tens are left.

(Yes)   No

tens	ones

tens	ones

3 | 5
− | 7

tens	ones
2	15

3̸ | 5̸
− | 7
  | 8

tens	ones
2	15

3̸ | 5̸
− | 7
2 | 8

I.

tens	ones

4 | 7
− | 9

Do you need to regroup?
Yes   No

tens	ones

3 | 7
− | 5

Do you need to regroup?
Yes   No

2.

tens	ones

7 | 4
− | 6

Do you need to regroup?
Yes   No

tens	ones

5 | 2
− | 8

Do you need to regroup?
Yes   No

Two-Digit Subtraction

Use Workmat 3 and base-ten blocks.
Circle **Yes** or **No** if you need to regroup. Subtract.

Step 1
Show 53.
Look at the ones. Do you need to regroup?

Step 2
Regroup 1 ten as 10 ones. Now there are 13 ones. Subtract 5 from 13. Write how many ones are left.

Step 3
Subtract the tens. Write how many tens are left.

(Yes)   No

tens	ones
5	3
− 1	5

tens	ones
4	13
5̸	3̸
− 1	5
	8

tens	ones
4	13
5̸	3̸
− 1	5
3	8

1.

tens	ones
3	2
− 1	6

Yes
No

tens	ones
4	4
− 2	5

Yes
No

tens	ones
5	0
− 3	8

Yes
No

2.

tens	ones
3	1
− 1	9

Yes
No

tens	ones
4	5
− 3	6

Yes
No

tens	ones
5	9
− 1	4

Yes
No

Practicing Two-Digit Subtraction

Step 1
Can you subtract the ones?

Step 2
Regroup if you need to. Subtract the ones.

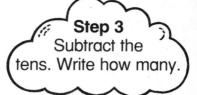

Step 3
Subtract the tens. Write how many.

```
   82
 - 46
```

```
  7 12
   8̸2̸
 - 46
    6
```

```
  7 12
   8̸2̸
 - 46
   36
```

Subtract. Regroup if you need to.

1.

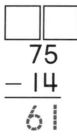

```
   75
 - 14
   61
```

```
   32
 - 23
```

```
   59
 - 42
```

```
   54
 - 39
```

2.

```
   49
 - 25
```

```
   22
 - 16
```

```
   64
 - 55
```

```
   33
 - 18
```

3.

```
   26
 - 19
```

```
   32
 - 12
```

```
   99
 - 26
```

```
   81
 - 24
```

Using Addition to Check Subtraction

Use Workmat 3 and base-ten blocks to solve.
Subtract. Regroup if you need to.
Cut and paste the addition problem that matches.

1.

$$
\begin{array}{r}
{}^{4}\;{}^{11} \\
5\!\!\!/1 \\
-\;15 \\
\hline
36
\end{array}
$$

$$
\begin{array}{r}
{}^{1} \\
36 \\
+\;15 \\
\hline
51
\end{array}
$$

$$
\begin{array}{r}
63 \\
-\;4 \\
\hline
\end{array}
$$

2.

$$
\begin{array}{r}
27 \\
-\;19 \\
\hline
\end{array}
$$

$$
\begin{array}{r}
75 \\
-\;66 \\
\hline
\end{array}
$$

3.

$$
\begin{array}{r}
46 \\
-\;27 \\
\hline
\end{array}
$$

$$
\begin{array}{r}
17 \\
-\;15 \\
\hline
\end{array}
$$

$\begin{array}{r}{}^{1}\\59\\+\;4\\\hline 63\end{array}$	$\begin{array}{r}{}^{1}\\19\\+\;27\\\hline 46\end{array}$	$\begin{array}{r}{}^{1}\\8\\+\;19\\\hline 27\end{array}$	$\begin{array}{r}{}^{1}\\36\\+\;15\\\hline 51\end{array}$	$\begin{array}{r}{}^{1}\\9\\+\;66\\\hline 75\end{array}$	$\begin{array}{r}2\\+\;15\\\hline 17\end{array}$

Problem Solving • Choose the Operation

Circle the problem that goes with the story. Then solve.

1. Sarah had 34 oranges. Her classmates ate 17 of them. How many oranges are left?

$$\begin{array}{r} 34 \\ + 17 \\ \hline \end{array}$$

_____ oranges

$$\begin{array}{r} \overset{2\ \ 14}{\cancel{34}} \\ - 17 \\ \hline 17 \end{array}$$

_17_ oranges

2. Jeff had 26 erasers. Kyle gave him 13 more. How many erasers are there in all?

$$\begin{array}{r} 26 \\ + 13 \\ \hline \end{array}$$

_____ erasers

$$\begin{array}{r} 26 \\ - 13 \\ \hline \end{array}$$

_____ erasers

3. Laura had 42 books. She gave 29 away. How many books are left?

$$\begin{array}{r} 42 \\ + 29 \\ \hline \end{array}$$

_____ books

$$\begin{array}{r} 42 \\ - 29 \\ \hline \end{array}$$

_____ books

4. Our class had 29 balls. Then, we got 12 more. How many balls are there in all?

$$\begin{array}{r} 29 \\ + 12 \\ \hline \end{array}$$

_____ balls

$$\begin{array}{r} 29 \\ - 12 \\ \hline \end{array}$$

_____ balls

Tally Tables

Count the shapes. Draw tally marks to show how many.

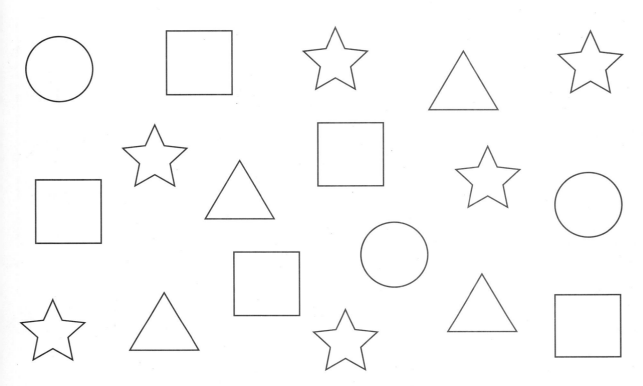

Shapes	
◯	││││
△	
▢	
☆	

I stands for 1.
卌 stands for 5.

Problem Solving • Use a Table

Draw tally marks for each crayon.

Favorite Color	
red	‖ ‖
blue	
green	
yellow	

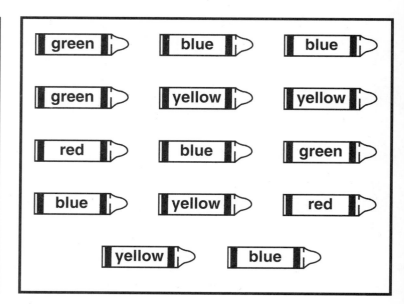

Circle the word to complete.

1.

_____ got four votes.   Green   (Yellow)

2.

_____ children like red best.   Two   Four

3.

_____ is the favorite color.   Red   Blue

4.

_____ is the least favorite color.   Green   Red

5.

_____ has one more vote than red.   Yellow   Green

Taking a Survey

Look around the room at your classmates.
Fill in the tally marks.
Then answer the questions.

1. What kind of shoes are your classmates wearing?

tennis shoes		
slip-ons		
boots		
other		

2. What color are your classmates' socks?

white		
blue		
striped		
other		

3. Are there more boots or tennis shoes in your room? _____

4. Are there fewer blue or white socks in your room? _____

Comparing Data in Tables

Read the sentences.
Fill in the missing tally marks to finish
the table for Mrs. Grigg's class.

Favorite Animals Ms. Hughes's class	
dog	ＨＨ l
cat	lll
bird	l
hamster	ll

Favorite Animals Mrs. Grigg's class	
dog	
cat	
bird	
hamster	

1. Four children in Mrs. Grigg's class like dogs.

2. Two more children in Mrs. Grigg's class like cats the best.

3. Birds got two more votes in Mrs. Grigg's class.

4. The two classes had the same number of votes for hamsters.

5. How many children in both classes like dogs the best?

_____

Picture Graphs

Use the graph to answer the questions.

Shapes	
circles	◯ ◯ ◯ ◯ ◯ ◯ ◯ ◯
triangles	△ △ △ △ △ △
squares	▢ ▢ ▢ ▢ ▢
trapezoids	⏢ ⏢ ⏢ ⏢

1. How many?

 ⑧   ____

△ ____    ____

2. Circle the shape that has more.

3. Circle the shape that has fewer.

4. How many more are there than ⏢ ?

____ more

Pictographs

Balloons	
red	
yellow	
blue	
green	
orange	

Each ⬭ stands for 5 balloons.

Put 5 counters on each balloon.
Use your graph to answer the questions.

I. How many?

red __20__   orange _____   green _____

blue _____   yellow _____

2. Circle the color that has most.

orange   blue

3. Circle the color that has fewest.

green   yellow

4. How many more blue are there than yellow?

_____

Horizontal Bar Graphs

Mr. Clark made a tally table of some things in his classroom. Color the graph to show how many items Mr. Clark counted.

Mr. Clark's Classroom	
clock	I
door	II
chalkboard	IIII
chairs	HHI HHI II
desks	HHI HHI I

Mr. Clark's Classroom												
clock												
door												
chalkboard												
chairs												
desks												

0 1 2 3 4 5 6 7 8 9 10 11 12

Problem Solving • Make a Graph

Ask 6 classmates to vote for their favorite zoo animal.
Fill in the tally table to show their answers.

Classmate's Name

1. _____

2. _____

3. _____

4. _____

5. _____

6. _____

Favorite Zoo Animal	
zebra	
lion	
elephant	

Use the tally table to fill in the graph.

Favorite Zoo Animal		
6		
5		
4		
3		
2		
1		
0		
zebra	lion	elephant

Certain or Impossible

Look at the box. Circle the pictures
that show what can come out of the box.

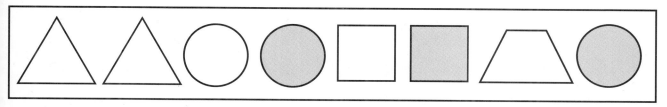

1.

2.

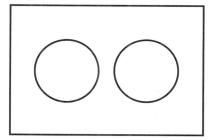

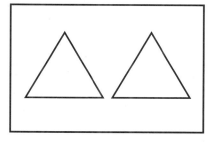

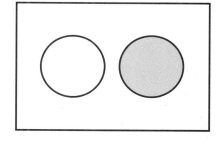

3.

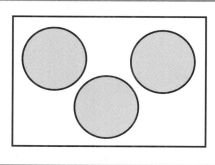

4.

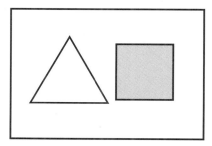

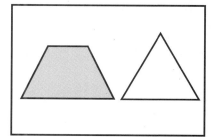

Interpreting Outcomes of Games

Color the spinners. Circle the color you think the spinner
will stop on most often. Tell a classmate why.

1.

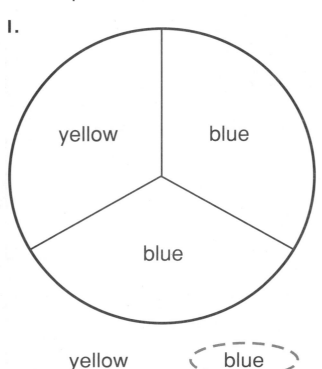

yellow   (blue)

2.

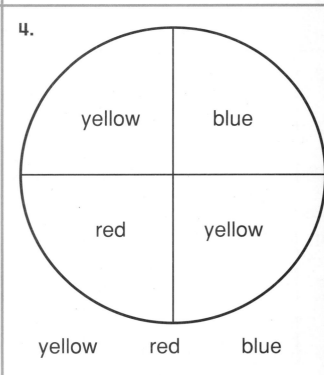

yellow   green

3.

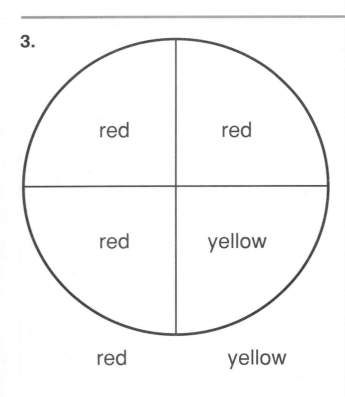

red   yellow

4.

yellow   red   blue

Most Likely

Put red counters and yellow
counters in a bag.
Pull out 1 counter.
Color the first box in the graph to
match the counter.
Put the counter in the bag.
Shake.
Do this 9 more times.
Write how many times
you pull out each color.

1. Use 3 red and 7 yellow counters.

Pull	1	2	3	4	5	6	7	8	9	10	____ red
Color											____ yellow

2. Use 1 red and 9 yellow counters.

Pull	1	2	3	4	5	6	7	8	9	10	____ red
Color											____ yellow

3. Use 2 red and 8 yellow counters.
Predict which color you will pull out more often.

	Red						Yellow				
Pull	1	2	3	4	5	6	7	8	9	10	____ red
Color											____ yellow

Less Likely

Color the spinner. Circle the color that you think the spinner
will stop on less often. Tell a classmate why.

1.

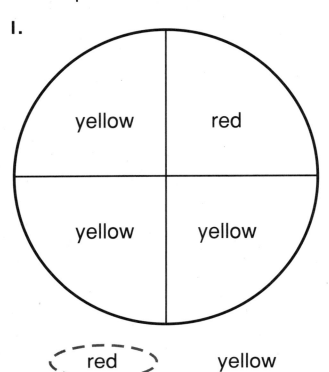

(red)   yellow

2.

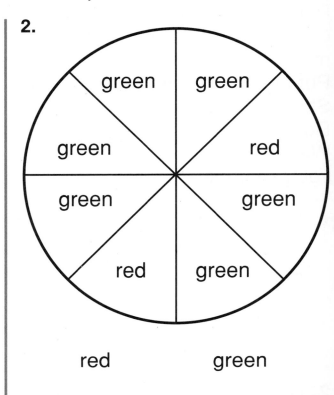

red   green

3.

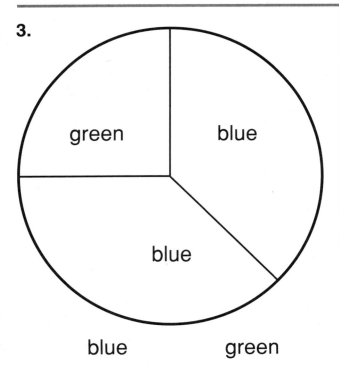

blue   green

4.

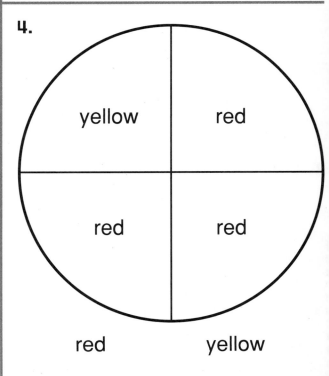

red   yellow

Identifying Solids

Cut out the objects.
Paste them to match the solid figures.

1.

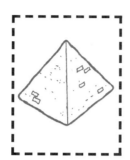 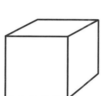 

2.

3.

    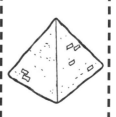 

Sorting Solid Figures

Use solid figures. Try each shape to see which
ones can stack, roll, or slide. Circle the correct figures.

1. It can roll.	2. It can stack.	3. It can slide.

Problem Solving • Look for a Pattern

Cut and paste the figure that belongs in the pattern.

1.

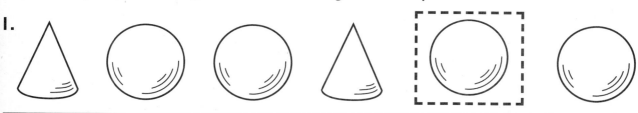

2.

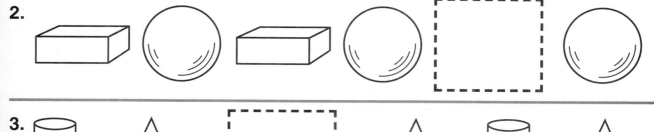

3.

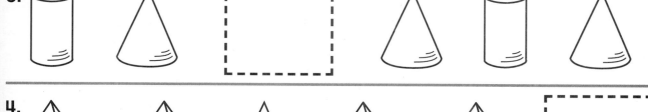

4.

5.

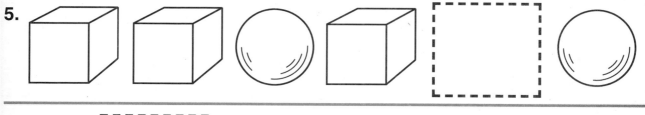

6.

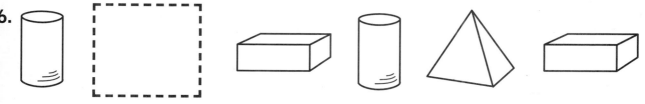

Making Plane Figures

Circle the plane figure that matches the
face of the solid figure.

I.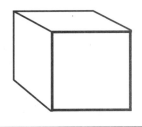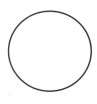

2.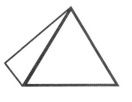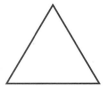

3.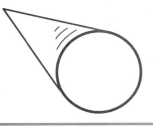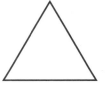

4.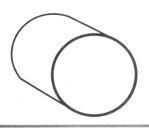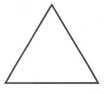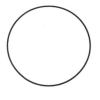

5.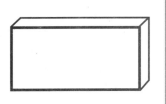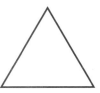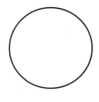

Name _____

Plane Figures

			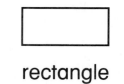
circle	square	triangle	rectangle

1. Color the rectangles.

How many? __3__

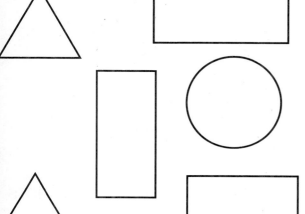

2. Color the circles.

How many? _____

3. Color the triangles.

How many? _____

4. Color the squares.

How many? _____

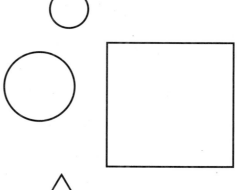

Sides and Corners

These figures have **sides.**

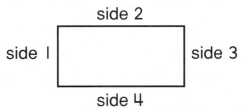

Sides meet at **corners.**

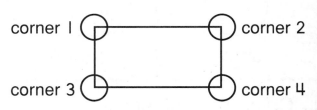

Trace the sides 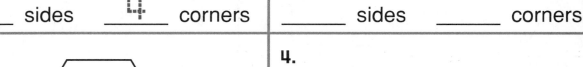 green .
Circle the corners red .
Write how many sides and corners.

1.

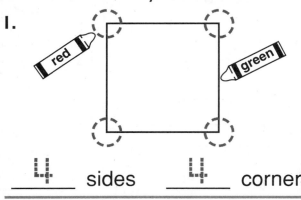

__4__ sides   __4__ corners

2.

_____ sides   _____ corners

3.

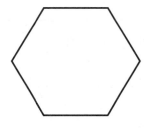

_____ sides   _____ corners

4.

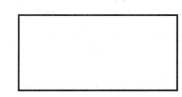

_____ sides   _____ corners

5.

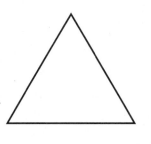

_____ sides   _____ corners

6.

_____ sides   _____ corners

Separating to Make New Figures

Trace the line or lines.
Write how many rectangles, triangles,
or squares you made.

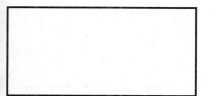

I rectangle

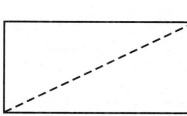

There are 2 triangles.

1.

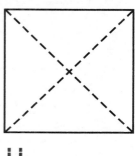

__4__ triangles

2.

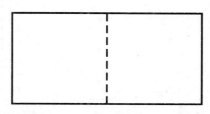

_____ squares

3.

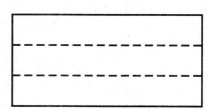

_____ rectangles

4.

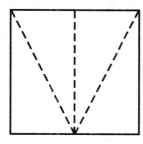

_____ triangles

5.

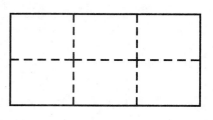

_____ squares

6.

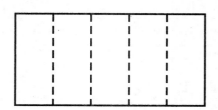

_____ rectangles

Congruent Figures

These figures are the same size and shape.

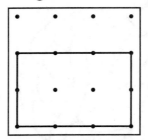 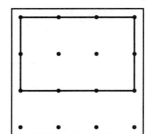 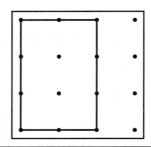

Circle the two figures that are the same size and shape.

1.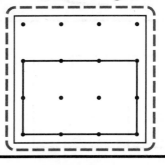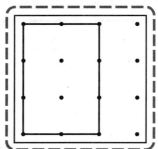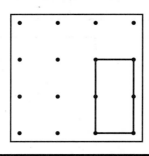

2.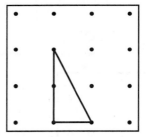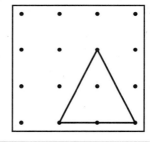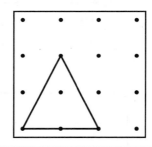

3.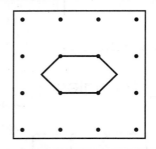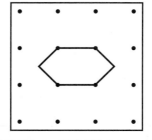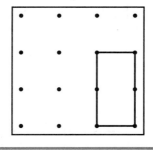

4.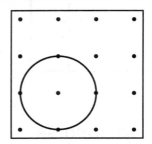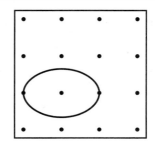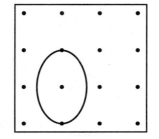

Line of Symmetry

The parts match.
It is a line of symmetry.

The parts do not match.
It is not a line of symmetry.

Color the picture that shows the line of symmetry.
Trace the line of symmetry.

1.
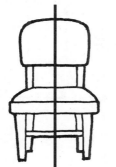

2.

3.
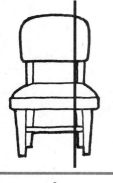

4.
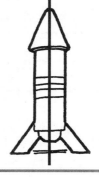

5.

6.

More Symmetry

Trace the line of symmetry.
Draw another line of symmetry where you can.

1.

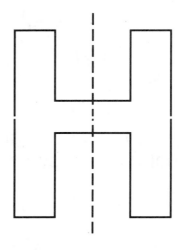

2.

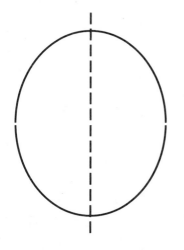

3.

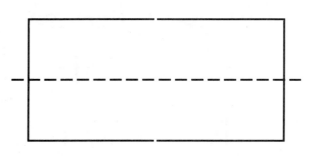

4.

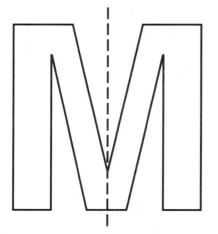

5.

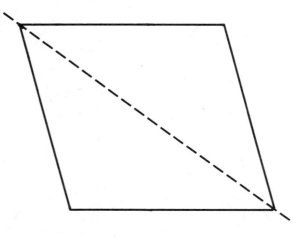

6.

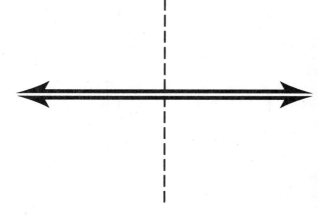

Moving Figures

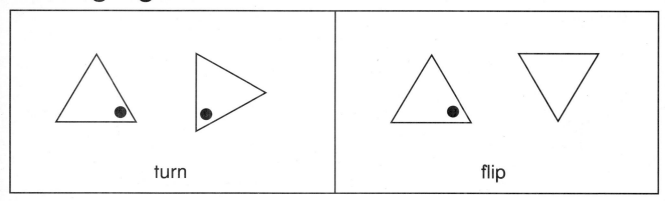

turn

flip

You will need pattern blocks. Place your block on the shape.
Move and trace your block to show a flip and turn.

1.

turn

2.

flip

3.

turn

4.

flip

5.

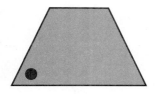

turn

6.

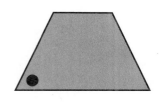

flip

More About Moving Figures

You will need attribute blocks.
Put your block on top of the figure.
Slide it to fit on top of the dashed one. Trace.

I.

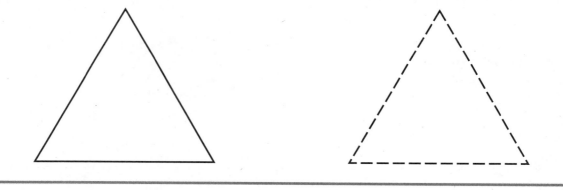

2.

3.

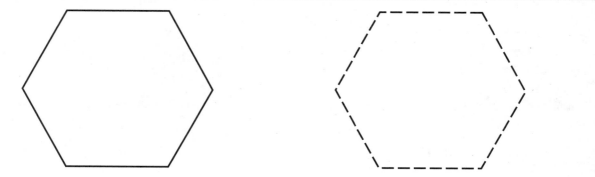

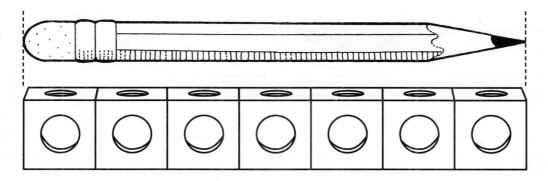

Using Nonstandard Units

This pencil is about 7 connecting cubes long.

Use connecting cubes to measure.
Circle the correct answer.

1.

about 3 cubes

(about 4 cubes)

2.

about 6 cubes

about 7 cubes

3.

about 4 cubes

about 5 cubes

4.

about 2 cubes

about 3 cubes

Measuring with Inch Units

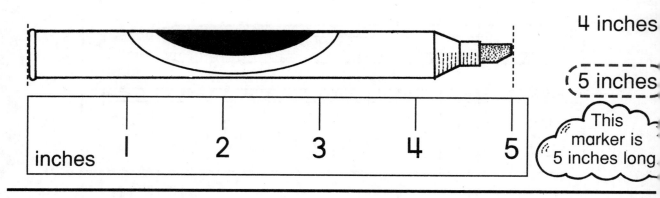

4 inches

(5 inches)

This marker is 5 inches long

Circle the length.

1.

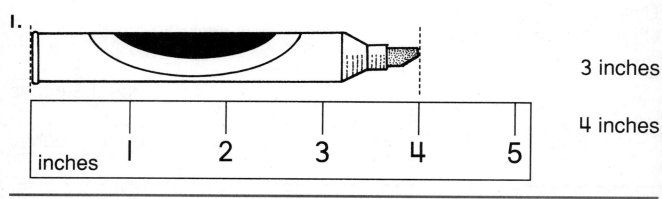

3 inches

4 inches

2.

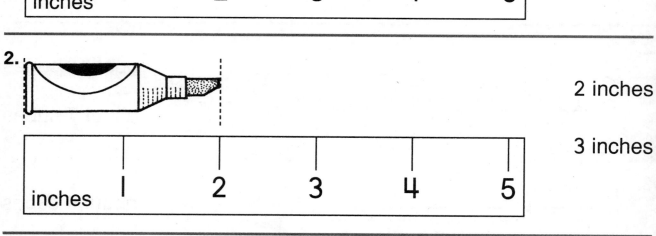

2 inches

3 inches

3.

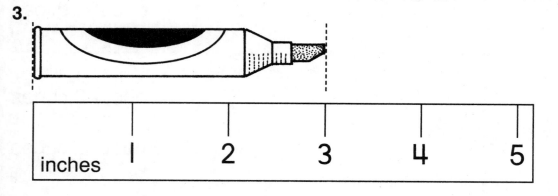

3 inches

4 inches

Name _____

Using an Inch Ruler

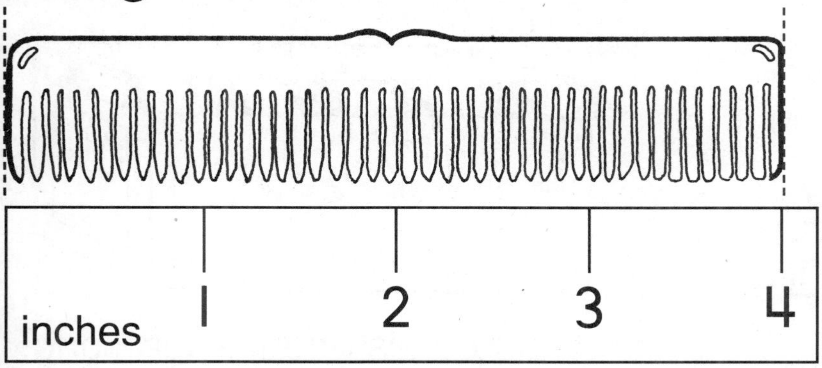

This comb is
4 inches long.

| inches | 1 | 2 | 3 | 4 |

___4___ inches

Use your inch ruler. Write the length of each object.

1.

_____ inches

2.

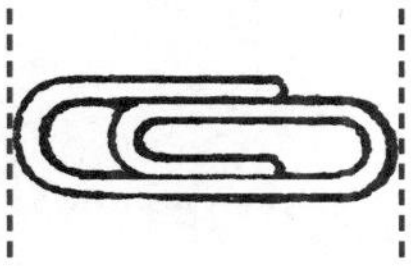

_____ inch

3.

_____ inches

4.

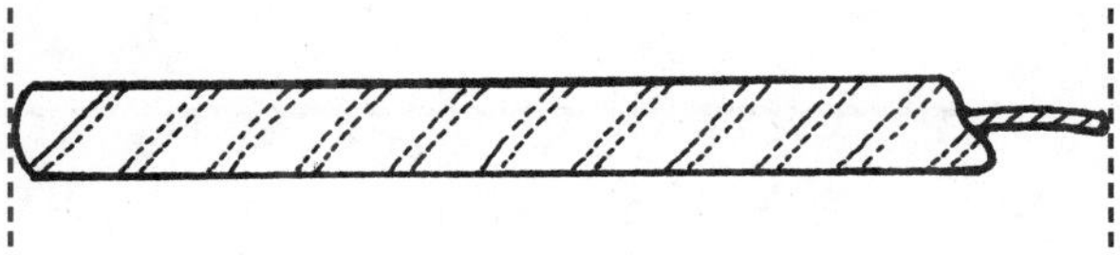

_____ inches

5.

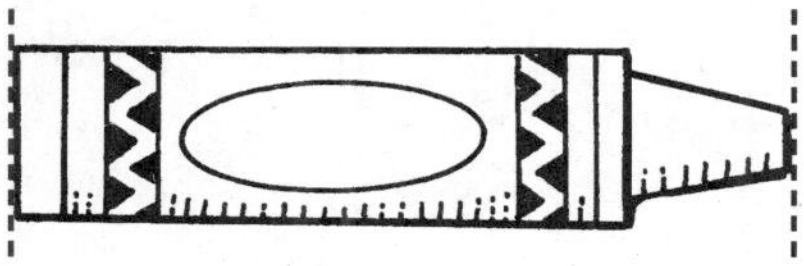

_____ inches

Foot

12 inches = 1 foot

A shoe box is about 1 foot long.

Find these things. Use a ruler to measure them.
Color [red ▷] the things that are less than 1 foot.
Color [blue ▷] the things that are more than 1 foot.

1.

book

2.

chair

3.

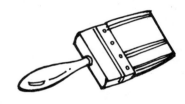

paintbrush

4.

shoe

5.

pencil

6.

child

7.

eraser

8.

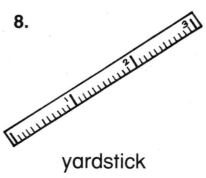

yardstick

9.

desk

Problem Solving • Guess and Check

Circle the better guess.
Then check your guess with a ruler.

		Check
1.	(about 3 inches) about 6 inches	__3__ inches
2.	about 4 inches about 2 inches	_____ inches
3.	about 2 inches about 4 inches	_____ inches
4.	about 4 inches about 5 inches	_____ inches
5.	about 1 inch about 3 inches	_____ inch

Centimeters

Circle the length.

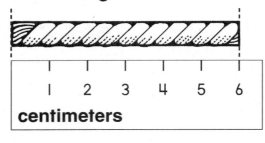

(6 centimeters)

This rope is
6 centimeters long.

8 centimeters

1.

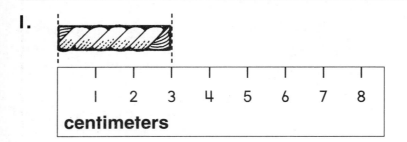

3 centimeters

1 centimeter

2.

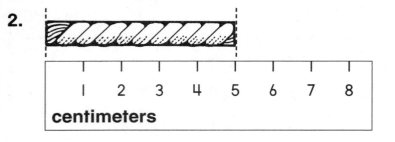

7 centimeters

5 centimeters

3.

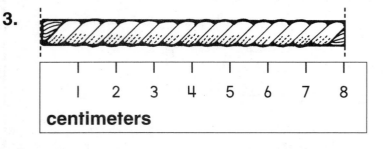

10 centimeters

8 centimeters

4.

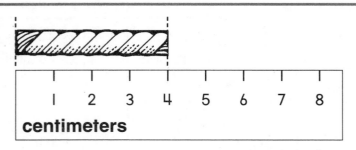

4 centimeters

2 centimeters

Decimeters

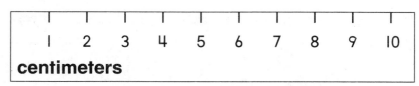

| 1 | 2 | 3 | 4 | 5 | 6 | 7 | 8 | 9 | 10 |

centimeters

1 decimeter = 10 centimeters

Use your centimeter ruler. Measure the objects below.
Circle the ones that are exactly 1 decimeter long.

1.

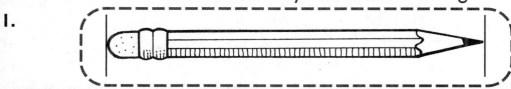

2.

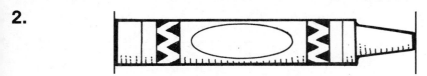

3.

4.

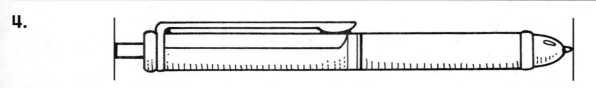

5.

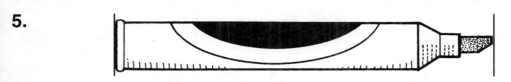

6.

7.

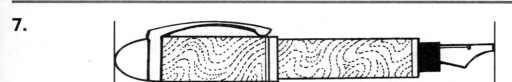

Name _____

LESSON
21.3

Exploring Perimeter

An ant walked 16 centimeters around the square.

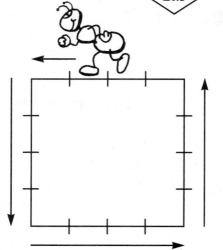

How far did the ant walk around each shape?
Count each space between the marks. Circle the correct answer.

1.

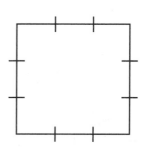

10 centimeters

(12 centimeters)

2.

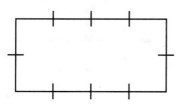

12 centimeters

14 centimeters

3.

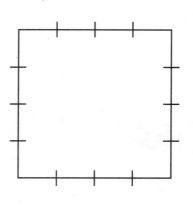

8 centimeters

16 centimeters

4.

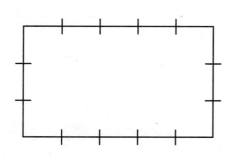

16 centimeters

14 centimeters

Problem Solving • Guess and Check

Guess how many squares will fit in the figure.
Write your guess. Use 1 inch squares. Put them over
the squares in the figure. Then count the squares to check.

1.

My guess is 5.

When I checked,
6 will fit.

Guess. _____ squares

Check. __6__ squares

2.

Guess. _____ squares

Check. _____ squares

3.

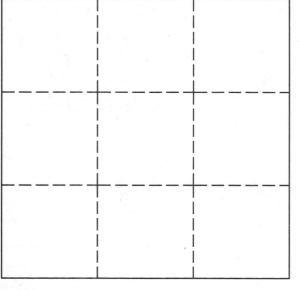

Guess. _____ squares

Check. _____ squares

4.

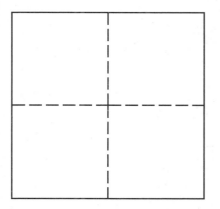

Guess. _____ squares

Check. _____ squares

Using Cups, Pints, and Quarts

I cup	I pint	I quart
	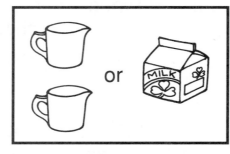	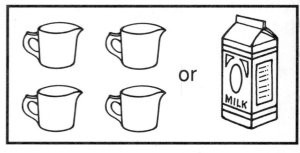
	2 cups = I pint	4 cups = I quart

Count the cups. Color the containers that hold the same amount.

I.

2.

3.

4.

5.

More and Less than a Pound

 less than I pound

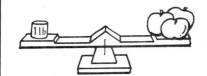 about I pound

 more than I pound

Find these objects.
Then look in the box.
Color the pictures to show what
the objects weigh.

less than I pound red
about I pound blue
more than I pound green

I.

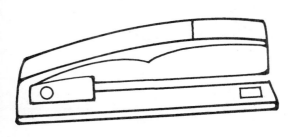

2.

3.

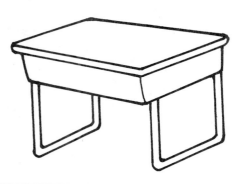

4.

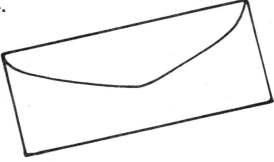

5.

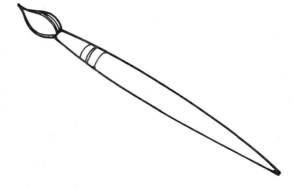

6.

Using a Thermometer

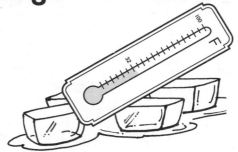

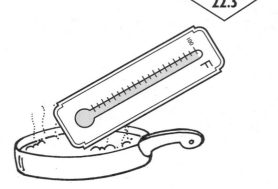

The thermometer line is low.
It shows a low temperature.
The ice is cold.

The thermometer line is high.
It shows a high temperature.
The water is hot.

Circle the better estimate of the temperature.

1.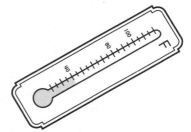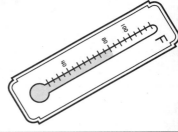

2.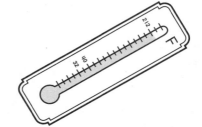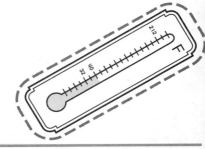

3.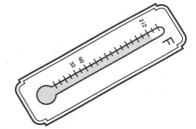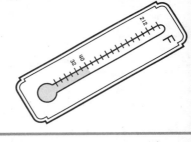

4.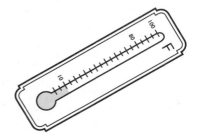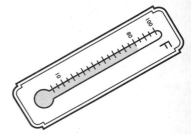

Choosing the Appropriate Tool

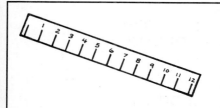		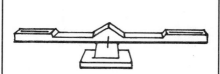
measures how long	measures how much	measures how heavy

Check the tool needed to answer the question.

		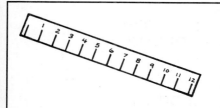	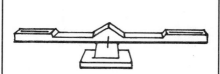
1. How much juice is in the jug?	✓		
2. How long is the rope?			
3. How much milk is in the mug?			
4. Which is heavier, an apple or an orange?			

Halves and Fourths

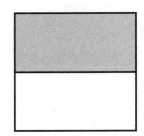

I part colored

2 equal parts

$= \dfrac{1}{2}$

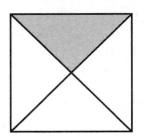

I part colored

4 equal parts

$= \dfrac{1}{4}$

Circle the fraction to show what part is colored.

1.

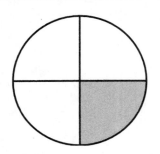

$\left(\dfrac{1}{2}\right)$ $\dfrac{1}{4}$

2.

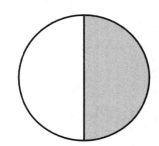

$\dfrac{1}{2}$ $\dfrac{1}{4}$

3.

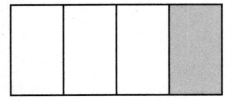

$\dfrac{1}{2}$ $\dfrac{1}{4}$

4.

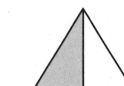

$\dfrac{1}{2}$ $\dfrac{1}{4}$

5.

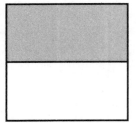

$\dfrac{1}{2}$ $\dfrac{1}{4}$

6.

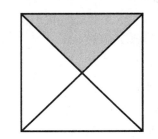

$\dfrac{1}{2}$ $\dfrac{1}{4}$

Thirds and Sixths

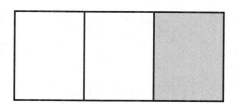

1 part colored

3 equal parts $= \dfrac{1}{3}$

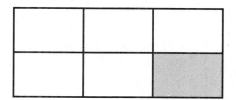

1 part colored

6 equal parts $= \dfrac{1}{6}$

Circle the fraction to show what part is colored.

1.

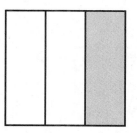

$\dfrac{1}{3}$ $\dfrac{1}{6}$

2.

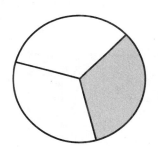

$\dfrac{1}{3}$ $\dfrac{1}{6}$

3.

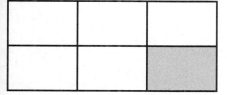

$\dfrac{1}{3}$ $\dfrac{1}{6}$

4.

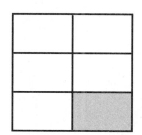

$\dfrac{1}{3}$ $\dfrac{1}{6}$

5.

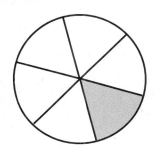

$\dfrac{1}{3}$ $\dfrac{1}{6}$

6.

$\dfrac{1}{3}$ $\dfrac{1}{6}$

More About Fractions

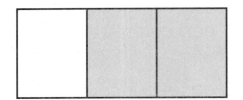

2 parts colored

3 equal parts

$= \dfrac{2}{3}$

Circle the correct fraction.

1.

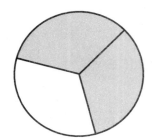

$\dfrac{1}{3}$ $\left(\dfrac{2}{3}\right)$

2.

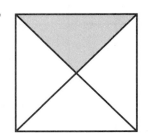

$\dfrac{1}{2}$ $\dfrac{1}{4}$

3.

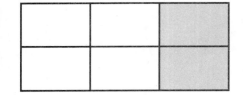

$\dfrac{2}{4}$ $\dfrac{2}{6}$

4.

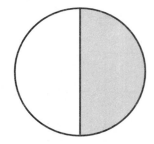

$\dfrac{3}{4}$ $\dfrac{2}{4}$

5.

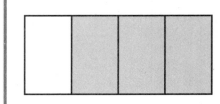

$\dfrac{1}{2}$ $\dfrac{2}{3}$

6.

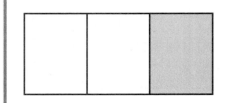

$\dfrac{2}{3}$ $\dfrac{1}{3}$

7.

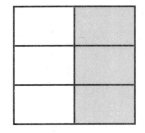

$\dfrac{3}{4}$ $\dfrac{3}{6}$

8.

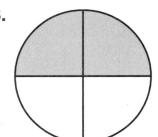

$\dfrac{2}{4}$ $\dfrac{2}{3}$

Parts of Groups

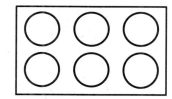

group of 6

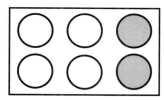

colored part is $\frac{1}{3}$

Circle the correct fraction.

1.

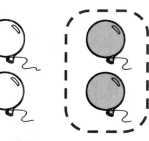

$\left(\frac{1}{2}\right)$ $\frac{1}{3}$

2.

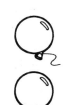

$\frac{1}{3}$ $\frac{1}{2}$

3.

$\frac{1}{4}$ $\frac{1}{3}$

4.

$\frac{1}{4}$ $\frac{1}{6}$

5.

$\frac{1}{2}$ $\frac{2}{3}$

6.

$\frac{1}{3}$ $\frac{2}{3}$

Problem Solving • Make a Model

Use fraction circles and the model to solve.
Complete the fraction.

I. Luke sliced a pie in fourths.
He ate 2 pieces and Betsy ate 2 pieces.
What part of the pie did Betsy eat?

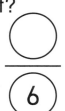

---

2. Jason and Sarah sliced a pie in sixths.
Jason ate 4 pieces. Sarah ate 2 pieces.
What part of the pie did Jason eat?

$$\frac{\bigcirc}{6}$$

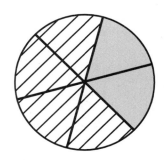

---

3. Laura sliced a pie in thirds.
Then she ate one piece.
What part of the pie did Laura eat?

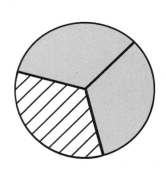

---

4. Simon sliced a pie in fourths.
Then he ate 3 slices.
What part of the pie did Simon eat?

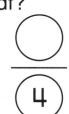

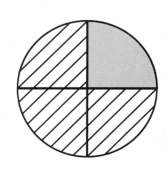

Name _____

Groups of Hundreds

Color to show how many.

1. 200
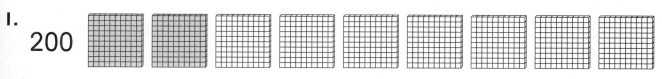

2. 500
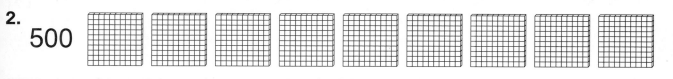

3. 900
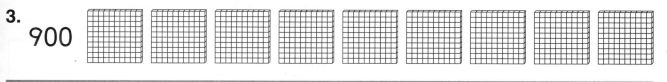

4. 700
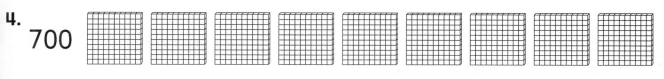

5. 300
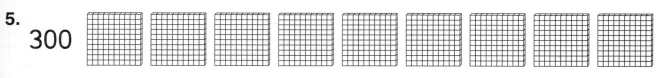

6. 800
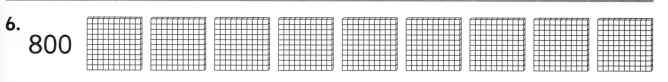

7. 100
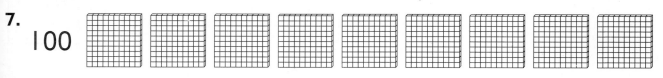

8. 400
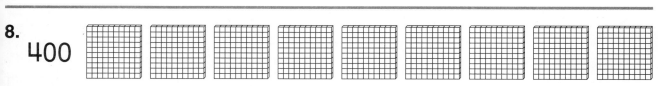

9. 600
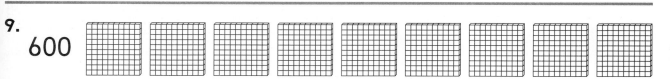

Numbers to 500

Write how many hundreds, tens, and ones.
Then write each number.

1.

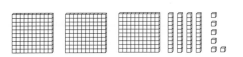

hundreds	tens	ones
3	4	6

$$3\,4\,6$$

2.

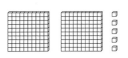

hundreds	tens	ones

3.

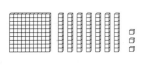

hundreds	tens	ones

4.

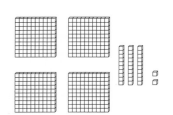

hundreds	tens	ones

5.

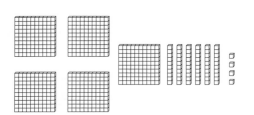

hundreds	tens	ones

Name _____

Numbers to 1,000

Write how many hundreds, tens, and ones.
Write each number.

1.

hundreds	tens	ones
2	3	6

236

2.

hundreds	tens	ones

3.

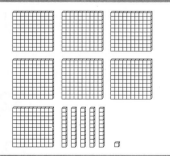

hundreds	tens	ones

4.

hundreds	tens	ones

5.

hundreds	tens	ones

Use a Model

Use base-ten blocks to make each model.
Circle the number.

1.

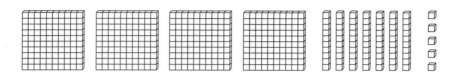

(475)

745

2.

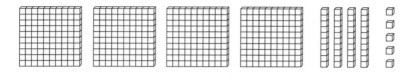

544

445

3.

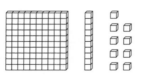

612

216

4.

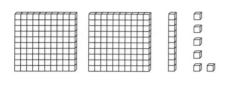

19

119

5.

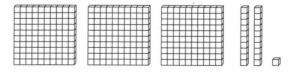

321

123

6.

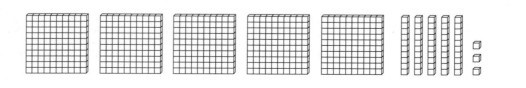

535

553

Building $1.00

Use coins. Show $1.00
with the coins named below.
Circle your answer.

$1.00 = 100 pennies
$1.00 = 100¢

I. half dollars

3 half dollars
(2 half dollars)

2. quarters

2 quarters
4 quarters

3. dimes

10 dimes
5 dimes

4. nickels

2 nickels
20 nickels

Greater Than

Compare the two models. Circle the number which is greater.

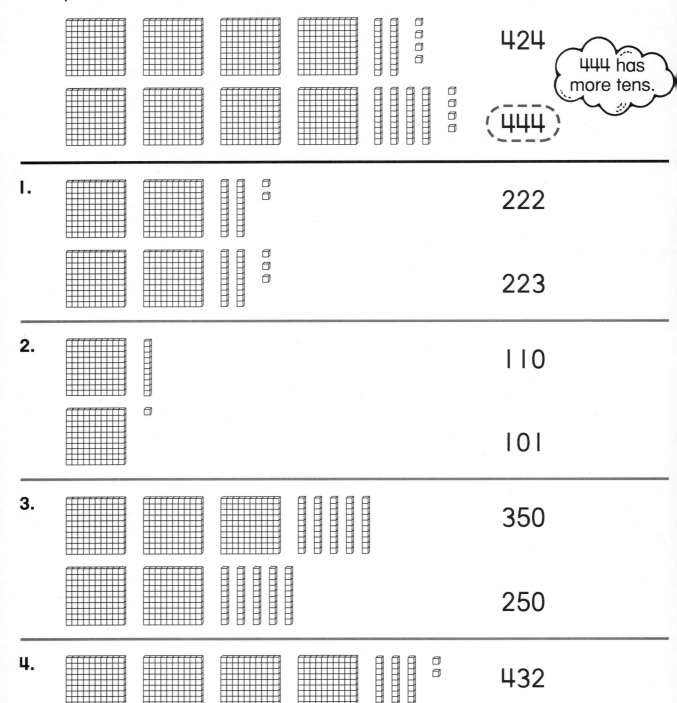

424

444 has more tens.

(444)

1. 222

223

2. 110

101

3. 350

250

4. 432

433

Less Than

Compare the two models. Circle the number which is less.

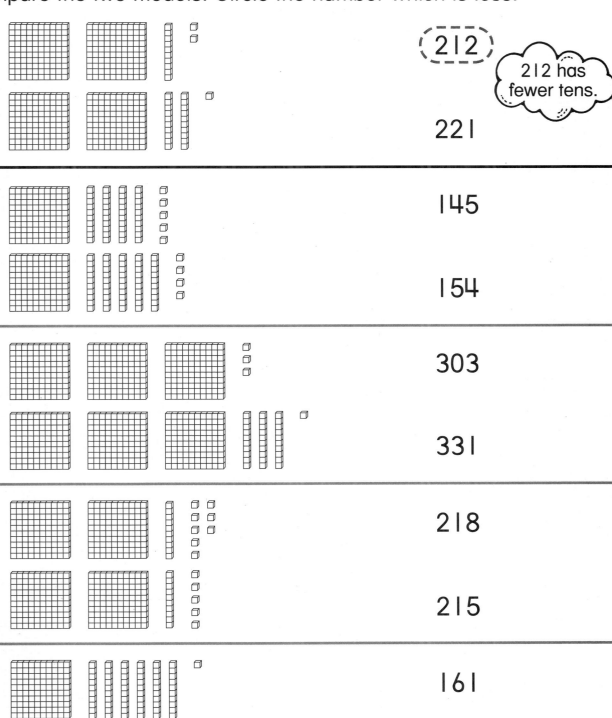

(212)

212 has fewer tens.

221

1. 145

    154

2. 303

    331

3. 218

    215

4. 161

    116

Greater Than and Less Than

Look at each model.
Complete the statement with **greater** or **less**.

1.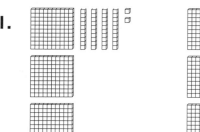

342 is _____**less**_____ than 432.

2.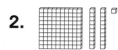

121 is **greater** than 111.

3.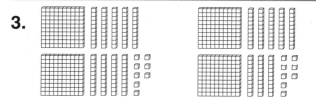

298 is _____ than 288.

4.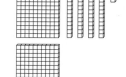

142 is _____ than 241.

5.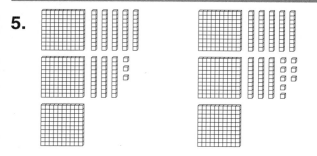

383 is _____ than 388.

6.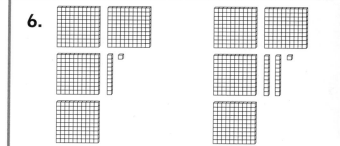

411 is _____ than 421.

7.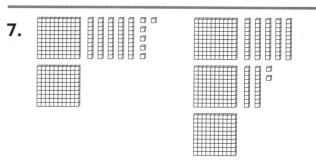

256 is _____ than 372.

8.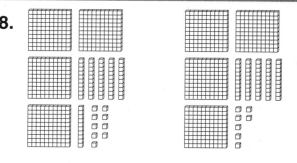

469 is _____ than 457.

Name _____

Before, After, and Between

LESSON 25.4

415 is just **before** 416.

416 is **between** 415 and 417.

417 is just **after** 416.

Write **before, after,** or **between** to complete the statement.

1.

319 is __**between**__ 318 and 320.

2.

222 is _____ 223.

3.

117 is _____ 116 and 118.

4.

545 is _____ 544.

TAKE ANOTHER LOOK R129

Name _____

Ordering Sets of Numbers

Use base-ten blocks to show each number.
Then write the numbers in order from least to greatest.

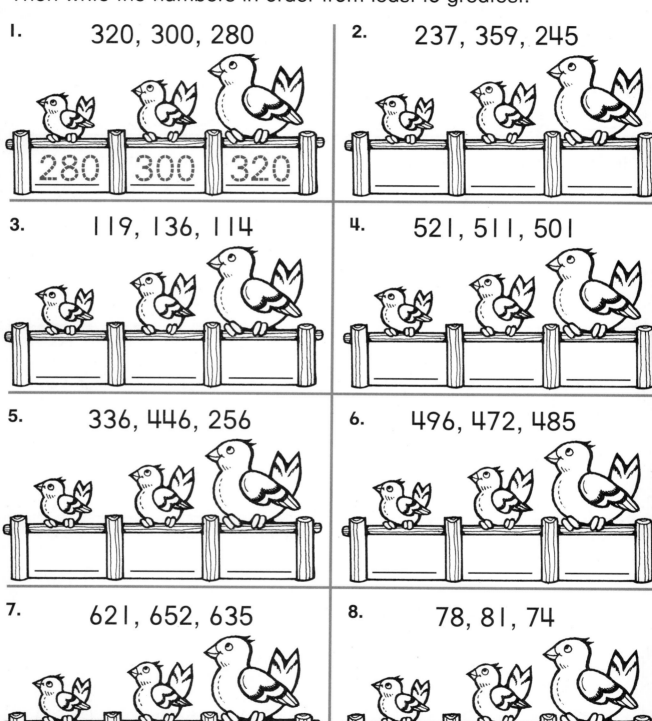

1. 320, 300, 280

280 300 320

2. 237, 359, 245

_____ _____ _____

3. 119, 136, 114

_____ _____ _____

4. 521, 511, 501

_____ _____ _____

5. 336, 446, 256

_____ _____ _____

6. 496, 472, 485

_____ _____ _____

7. 621, 652, 635

_____ _____ _____

8. 78, 81, 74

_____ _____ _____

Name _____

Modeling Addition of Three-Digit Numbers

Circle 10 ones and draw 1 ten to show regrouping.
How many ones now? Write that number.
Write how many tens and hundreds.

1.

hundreds	tens	ones
	1	
1	4	4
+2	4	7
3	9	1

2.

hundreds	tens	ones
1	2	6
+5	6	7

3.

hundreds	tens	ones
2	1	9
+2	3	2

4.

hundreds	tens	ones
1	0	5
+1	3	6

Adding Three-Digit Numbers

Step 1
Add ones. Regroup
if you need to. Write
the number of ones.

Step 2
Add tens. Regroup 11 tens
as 1 hundred and 1 ten.
Write the number of tens.

Step 3
Add hundreds.
Write the number
of hundreds.

hundreds	tens	ones
☐	☐	
1	4	7
+2	7	1
		8

hundreds	tens	ones
1	☐	
1	4	7
+2	7	1
	1	8

hundreds	tens	ones
1	☐	
1	4	7
+2	7	1
4	1	8

Use base-ten blocks and Workmat 5. Add.

1.

hundreds	tens	ones
1	☐	
3	3	6
+2	7	1
6	0	7

2.

hundreds	tens	ones
☐	☐	
1	2	9
+3	9	0

3.

hundreds	tens	ones
☐	☐	
2	6	4
+2	6	2

4.

hundreds	tens	ones
☐	☐	
3	8	7
+2	3	1

5.

hundreds	tens	ones
☐	☐	
1	7	2
+4	6	1

6.

hundreds	tens	ones
☐	☐	
4	2	0
+1	9	2

Modeling Subtraction of Three-Digit Numbers

Do you need more ones?
If so, cross out 1 ten and draw
10 more ones.
Then cross out to subtract the
ones, tens, and hundreds.
Write how many are left.

$$
\begin{array}{r}
3 \quad 6 \quad 2 \\
-1 \quad 2 \quad 4 \\
\hline
\end{array}
$$

1.

hundreds	tens	ones
	5	12
3	6̶	2̶
− 1	2	4
2	3	8

hundreds	tens	ones

2.

hundreds	tens	ones
	□	□
2	5	2
− 1	3	5

hundreds	tens	ones

3.

hundreds	tens	ones
	□	□
8	6	1
− 4	0	7

hundreds	tens	ones

Subtracting Three-Digit Numbers

Subtract the ones. Do you need more tens?
If so, cross out 1 hundred and draw 10 more tens.
Then cross out to subtract the tens and hundreds.
Write how many are left.

1.

hundreds	tens	ones
3	16	
4	6	5
− 2	7	3
1	9	2

hundreds	tens	ones

2.

hundreds	tens	ones
6	2	9
− 4	8	0

hundreds	tens	ones

3.

hundreds	tens	ones
8	4	7
− 4	5	6

hundreds	tens	ones

Adding and Subtracting Money

1. How much would both sailboats cost?

$3.85
+ 2.10
$5.95

2. How much more does the large sailboat cost than the small sailboat?

$3.85
– 2.10
$.

3. How much would both kites cost?

$4.73
+ 2.33
$.

4. How much more does the large kite cost than the small kite?

$4.73
– 2.33
$.

Adding Equal Groups

Use counters to show equal groups.
Draw them. Write how many in all.

1.

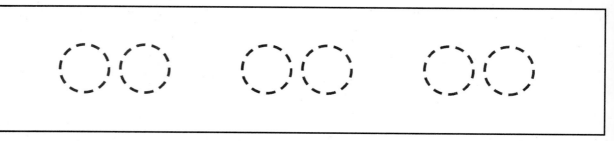

3 groups of 2 $2 + 2 + 2 = \underline{6}$

2.

4 groups of 3 $3 + 3 + 3 + 3 = \underline{}$

3.

3 groups of 5 $5 + 5 + 5 = \underline{}$

Multiplying with 2 and 5

Use counters to show equal groups.
Draw them. Write how many in all.

1.

2 groups of 2

$2 + 2 =$ _4_

$2 \times 2 =$ _4_

2.

4 groups of 2

$2 + 2 + 2 + 2 =$ ____

$4 \times 2 =$ ____

3.

3 groups of 5

$5 + 5 + 5 =$ ____

$3 \times 5 =$ ____

Multiplying with 3 and 4

Draw 4 scoops of ice cream on each ice cream cone.
Fill in the missing numbers.

1.

How many groups of 4?

_2_

_2_ × 4 = _8_

2.

How many groups of 4?

____

____ × 4 = ____

3.

How many groups of 4?

____

____ × 4 = ____

4.

How many groups of 4?

____

____ × 4 = ____

Problem Solving • Draw a Picture

Choose the multiplication sentence.

1. There are 5 boats on the lake. Each boat has 2 sails. How many sails in all do the 5 boats have?

$(5 \times 2 = 10)$

$5 \times 3 = 15$

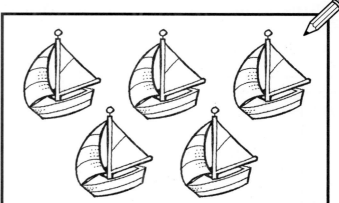

2. There are 4 puppies in the box. Each puppy has 3 spots. How many spots in all do the 4 puppies have?

$4 \times 3 = 12$

$2 \times 3 = 6$

3. There are 3 babies in the room. Each baby is holding 2 blocks. How many blocks in all do the 3 babies have?

$4 \times 2 = 8$

$3 \times 2 = 6$

4. There are 2 boxes on the shelf. Each box has 4 turtles in it. How many turtles in all do the 2 boxes have?

$2 \times 6 = 12$

$2 \times 4 = 8$

How Many in Each Group?

Use counters. Make equal groups in the boxes.
Draw them. Write how many are in each group.

1. Use 9 counters.

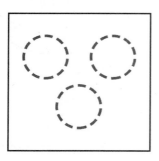 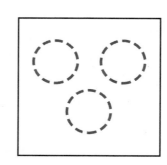 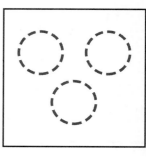

How many are in each group? _____3_____

---

2. Use 12 counters.

 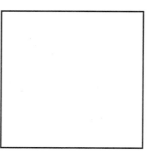  

How many are in each group? _____

---

3. Use 10 counters.

How many are in each group? _____

How Many Equal Groups?

Use counters. Put them in equal groups.
Draw them. Write how many groups.

I. Use 9 counters. Put them in groups of 3.

_____3_____ groups

---

2. Use 12 counters. Put them in groups of 6.

_____ groups

---

3. Use 8 counters. Put them in groups of 2.

_____ groups

Problem Solving • Draw a Picture

Read the problem. Finish each picture.

1. There are 9 hearts. There are 3 equal groups of hearts. How many are in each group?

____3____ in each group

2. There are 10 stars. There are 5 equal groups of stars. How many are in each group?

_____ in each group

3. There are 15 triangles. There are 3 triangles in each group. How many groups?

_____ groups

4. There are 12 squares. There are 4 in each group. How many groups?

_____ groups

Problem Solving • Choose a Strategy

Match the picture to the problem by writing
the letter under the problem. Solve.

1. There are 6 baskets.
Each basket had 2 eggs
in it. How many eggs were
there in all?

_____D_____

A.

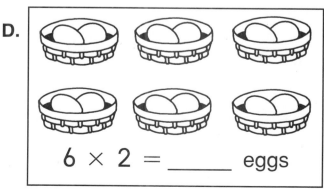

$17 + 8 = $ _____ flowers

2. Jay, Steve and Chris went
to the fair. They each
spent $4.00. How much
money did they spend
in all?

_____

B.

_____ pears each

3. Ashley gave 9 pears to
3 friends. She gave an
equal number to each.
How many pears did each
friend get?

_____

C.

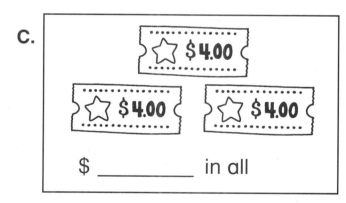

$4.00 $4.00 $4.00

$ _____ in all

4. Pam's garden had 17
flowers. 8 more flowers
bloomed. How many
flowers bloomed in all?

_____

D.

$6 \times 2 = $ _____ eggs